Reproduction in Flowering Plants

Reproduction in Flowering Plants

Author

Miss. Tejashri. C. H

M.Sc., B.Ed.

Lecturer
Department of Botany
Sri. Channabasaveshwar Post Graduate Studies
& Research Center Bhalki – 585328

ACADEMIC
ENCLAVE

Edited By

Mr. Juber Hawaldar (M.Sc., B.Ed.)
Lecturer, Department of Physics
Sri. Channabasaveshwar Post Graduate Studies
& Research Center Bhalki – 585328

Proof Reading

Prof. Meenakumari Patil ***(M.Com. M.Phil. M.A., M.Ed.)***
Principal & Assistant Professor
Department of Commerce
Sri. Channabasaveshwar Post Graduate Studies
& Research Center Bhalki – 585328

Miss. Ashmita Nirgude ***(M.Sc., B.Ed.)***
Head, Department of Botany
Sri. Channabasaveshwar Post Graduate Studies
& Research Center Bhalki – 585328

Mr. Pashusab N ***(M.Sc., B.Ed.)***
Head, Department of Physics
Sri. Channabasaveshwar Post Graduate Studies
& Research Center Bhalki – 585328

Mr. Arvinda N ***(M.Sc.)***
Lecturer, Department of Botany
Sri. Channabasaveshwar Post Graduate Studies
& Research Center Bhalki – 585328

First Edition, 2024

ISBN: 978-81-971898-5-2 (Paperback)

Published by
ACADEMIC ENCLAVE
House No. 10/15, Ground Floor,
Block-A, Street No. 14/3, Kamal Vihar,
Kamal Pur, Burari, Delhi-110084
Email: info@academicenclave.com
Mobile: +91-7042423166
Website: www.academicenclave.com

Dedication

This Book is dedicated to my Parents because of their efforts and Sacrifice I became an educated and written this book.

This Book is dedicated to my teachers, who gave me knowledge about alphabets and enhanced my skills.

This Book is dedicated to my Family, Relatives and Friends who encourages me at every time. Also, this Book dedicated to my all Supporters and Well-wishers.

FOREWORD

The remarkable world of flowering plants, with their diverse reproductive strategies, serves as a cornerstone of ecological balance and agricultural productivity. This Book explores the intricacies of reproduction in these plants, focusing on essential processes such as pollination, fertilization, and seed development.

Understanding these mechanisms is crucial not only for botanists but also for anyone interested in the interconnectedness of life on Earth. As flowering plants contribute significantly to our ecosystems and food systems, studying their reproductive strategies provides insights into biodiversity and conservation efforts.

The author, **Miss.Tejashri. CH**, has crafted this text with a deep understanding of both the subject matter and the pedagogical challenges involved in teaching it. With clarity and precision, **Miss. Tejashri. CH** presents the core concepts of Reproduction in Flowering Plants, breaking down complex ideas into manageable and comprehensible segments.

This work aims to serve as a valuable resource for students, educators, and nature enthusiasts alike. It reflects the collective efforts of those who inspired and supported this exploration, and I hope it fosters a greater appreciation for the beauty and complexity of flowering plants.

With anticipation for the discoveries that await you,

Sincerely,

Dr. Shivanand Sakpal

M.Sc., B.Ed., Ph.D.

Head of the Department & Asst. Professor

Department of Botany

Sri. Channabasaveshwar Art, Commerce &

Science Degree College, Bhalki- 585328

ACKNOWLEDGMENTS

First and foremost, I want to express my heartfelt gratitude to the dedicated team at **Academic Enclave** for their unwavering support and commitment to this Book. Their guidance, patience, and tireless efforts have been instrumental in turning the vision of this book into a reality.

I also want to express my heartfelt gratitude to the Principal and Staff members of "Sri Channabasaveshwar Post Graduate Studies and Research Centre Bhalki" for their valuable suggestions and support to this book.

I would like to extend my heartfelt thanks to my mentors and colleagues, whose insightful feedback and encouragement have been invaluable. Your expertise and constructive criticism have not only helped refine the content of this book but have also inspired me to approach reproduction in flowering plants with greater depth and clarity.

I also want to express my heartfelt gratitude to Dr. Santosh Kumar Teerthe, Dr. Shivanand Sakpal, Dr. Kauthale Ramling, Prof. Meenakumari Patil, Mr. PashuSab N, Mr. Vishwajeet M, Mrs. Jyothi D. Anandwade, Mr. Arunkumar Mahajan, Miss. Anjali Kale, Miss. Ashmita Nirgude, Miss. Ragini, Mr. Sharanappa, Mr. Arvind.S, Mrs. Anjali Kawthe, Mr. Mahadev Borale, Miss. Dhanshrree. CH, Mr. Nagabhooshan. S for their valuable suggestions to this book.

I am particularly indebted to my students, whose curiosity and enthusiasm have continually motivated me to explore new ways of presenting complex ideas. Your questions and insights have played a crucial role in shaping the approach and examples found in these pages.

Finally, I would like to thank you, the reader, for choosing to embark on this exploration of Reproduction in Flowering plants. It is my hope that this book serves as a valuable resource and inspires a deeper appreciation for the beauty and relevance of the principles that govern our biological world.

Miss. Tejashri.CH

PREFACE

It gives me great pleasure in bringing out the First Edition of this book.

Welcome to **Reproduction in Flowering Plants** unveils a fascinating world of diversity and adaptation. As the most diverse group of plants on Earth, angiosperms play a crucial role in our ecosystems, influencing everything from food production to habitat stability. Their reproductive strategies are intricate and varied, reflecting millions of years of evolution.

This book aims to explore the essential processes of reproduction in flowering plants, from the structure of flowers and mechanisms of pollination to the complex events of fertilization and seed development. By examining these processes, we gain insight into how plants interact with their environment and adapt to changing conditions.

Understanding these reproductive mechanisms is not only vital for botanists but also for anyone invested in agriculture, conservation, and environmental sustainability. As we face pressing global challenges, such as climate change and habitat loss, the knowledge contained within these pages can help us appreciate the importance of preserving plant diversity.

I invite you to embark on this journey into the world of flowering plants, where this chapter reveals the wonders of nature's design. Together, let us deepen our understanding of these remarkable organisms and their vital contributions to our planet.

I look forward for the suggestions from the readers for the improvement of the book. Sincerely,

Miss. Tejashri.CH

CONTENTS

1

REPRODUCTION IN ORGANISM

1.1. Reproduction

"It is biological process by new individual organisms or offsprings are produced from their parent or parents." Or "Reproduction is a biological process by which organisms produces exact copies of its own structures that are biologically and genetically similar to the parent or parents".

1.2. Types of Reproduction

- **Vegetative reproduction:** It is method of reproduction in which we can able produce new individuals from any parts of the organism particularly in plants. This type of production is usually takes place in plants but it does not involve the fusion or involvement of any gametes.
- **Asexual reproduction:** It is a type of reproduction in which only single organism gives rise to a new individual, it does not involve the fusion of gametes. These may include different methods such as **Binary fission, Budding, Fragmentation** and **Sporogenesis.** This kind of reproduction is take place mostly in Bacteria and Fungi.
- **Sexual reproduction:** It is type of reproduction in that involves the production of offspring by the fusion of male and female gametes.

The sexual reproduction take place usually in organism which contains gametes but, in this book, particularly we deal with the reproduction in plants that is the sexual reproduction in Flowering plants.

2

SEXUAL REPRODUCTION IN FLOWERING PLANTS

2.1 Flower

"Flower is a modified shoot meant for sexual reproduction."

2.2 Whorls or Parts of Flower

There are typically four whorls or parts of modified leaves such as Calyx, Carolla, Androecium and Gynoecium.

SI.No.	Whorl name	Unit name	
1	Calyx	Sepal	Accessory whorls
2	Carolla	Petal	
3	Androecium	Stamen	Essential or Reproductive whorls
4	Gynoecium	Gynoecium or Carpel	

2.3 Diagrammatic Representation of Flower

A The flower is typically consisting of different parts which are as fallows

- Stalk of the flower is called **Pedicel.**
- The whorls of the flower are borne on the expanded tip called **Receptacle or Thalamus.**
- The outermost whorl, **Calyx (Sepal)** which protects the flower in the bud.
- The next whorl, **Carolla (Petal)** which attracts insects and birds for pollination.
- The flowers are morphological and embryological marvels and the sites of sexual reproduction.

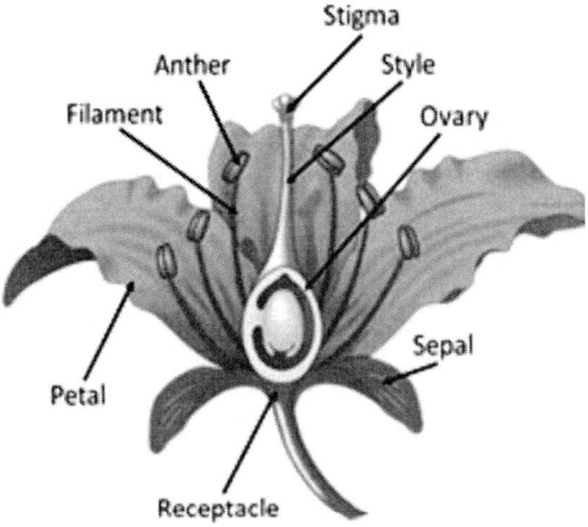

Fig. 2.1: Structure of Flower

2.4 Perianth

A flower in which calyx and corolla are undifferentiated is called perianth. Such flower is called homochlamydous flower. A unit or segment of perianth is called tepal.

Ex: Champak flower, Coconut, Onion etc.

Q1. Name the parts of an angiosperm flower in which development of male and female gametophyte takes place.

Answer: Inside the anther, the cells of micro sporangia develop as male gamete. Inside the ovary megasporangial cells develop as female gametes.

3

PRE-FERTILIZATION EVENTS

- It involves the structure and development of Male reproductive structure (Androecium).
- It also involves structure and development of female reproductive structure (Gynoecium).

3.1 Stamen and its Structure

Stamen is the male reproductive structure of flower and it contains different parts.

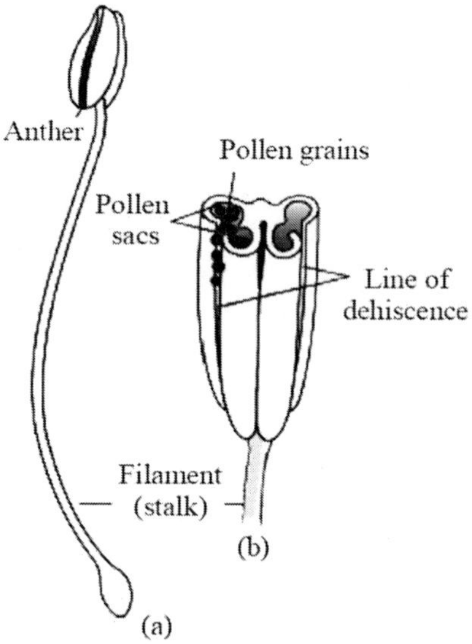

Fig. 3.1: Structure of Stamen

- A typical stamen has a slender stalk called **Filament.**
- The terminal portion generally has bilobed club shaped structure called **Anther.**
- A typical angiosperm anther is **bilobed** with each lobe having two **thecae** i.e. they are **Dithecous.**

- The anther is a four-sided (tetragonal) structure consisting of four **microsporangia** located at the corner, two in each lobe.
- The **microsporangia** develop further and become **pollen sacs.**
- **Pollen sacs** are packed with **pollen grains**.

3.2. Microsporangium

- In a **T.S** it appears almost circular in outline.
- It is surrounded by four wall layers called **epidermis, endothecium, middle layer** and **tapetum**.
- The outer three wall layers perform the function of protection and help in dehiscence of anther to release pollen grains.
- The innermost layer **tapetum** nourishes the developing pollen grain.
- Cells of tapetum possess dense cytoplasm and have more than one nucleus.
- In young **anther, microsporangium** contains **sporogenous tissue** at the centre which undergoes meiosis to produce **microspore tetrads** which further develop into pollen grains.

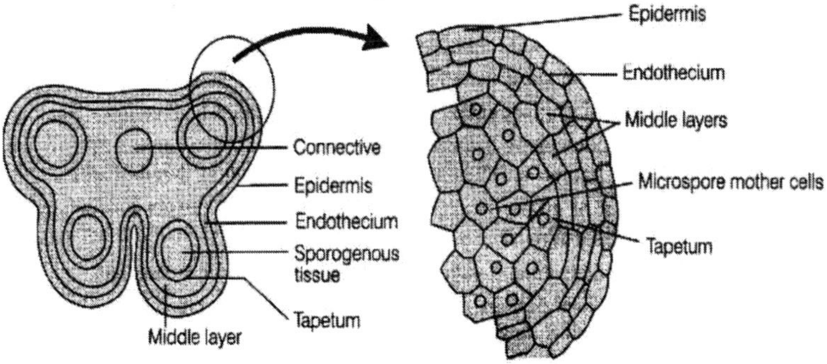

Fig. 3.2: structure of Microsporangium (T.S of young anther)

3.3 Tapetum

It is innermost layer of the anther wall and is developed from primary parietal cells. It is present below the middle layer. The cells of tapetum are large, radially elongated multinucleated and contain dense cytoplasm.

3.3.1 Types of Tapetum

Based on its behaviour the tapetum is of two types.

- **Glandular Tapetum:** It is also known as **parietal tapetum or secretory tapetum**. In this type the tapetal cells remain intact in their original position throughout the development of pollen grains or microspore.

  Glandular tapetum is more common in Angiosperms then the amoeboid tapetum. The tapetal cells do not degenerates. The cells are rich in protoplasmic contents and they secrete a nutritive secretion from their inner surface. This secretion is used by pollen grains. The tapetal cells contains **Ubisch bodies** these bodies contribute a material known as sporopollenin which involve in the thickening of exine of pollen grains.

- **Amoeboid Tapetum (Periplasmodial):** It is also known as **Periplasmodial tapetum**, in this tapetum the inner and radial walls of the tapetal cells breakdown and protoplast become a plasmodial mass called tapetal periplasmodium. The pollen mother cells obtain nourishment from it for pollen development. This type of tapetum is found in anthers of hydrophytes (Ex: Typha).

3.3.2 Role or Function of Tapetum

- It provides nutrition for the developing pollen grains.
- It plays an important role in the pollen wall formation.
- It Ubisch bodies present in the tapetal cell contribute a material called sporopollenin, which is involved in the formation Exine of pollen grains.
- The plastids of the tapetal cells synthesis material like pollen Kitt and typhine etc. these materials protect the pollen from Uv rays and also helpful in insect dispersal of pollen grains.

- The cells of tapetum releases an enzyme called callose, which degrades the callose around the pollen tetrads and hence the pollen grains in the tetrads are separated.
- Tapetum is essential for the formation of viable pollen grain.

3.4 Microsporogenesis

"The process of formation of pollen grains or microspore from pollen mother cell (PMC) through meiosis is called Microsporogenesis."

- Each cell of a sporogenous tissue is called **pollen mother cell (PMC) or Microspore mother cell (MMC).**
- As the anther matures, the cells of sporogenous tissue undergo meiosis to form **microspore tetrad (n).**
- The microspore, as they are formed are arranged in a cluster of four cells hence the name **microspore tetrad or pollen tetrad.**
- As the anthers mature the pollen tetrads dissociate from each other and develop into **pollen grains.**
- Inside each microsporangium several thousand of microspores or pollen grains are formed, that are released with the dehiscence of anther.

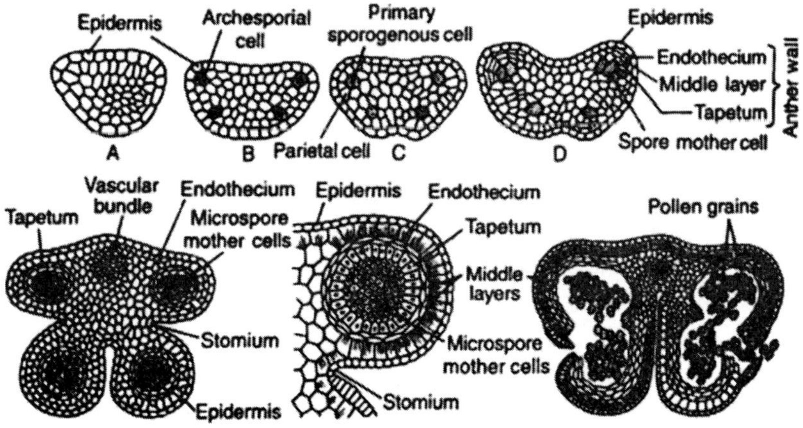

Fig. 3.3: Complete Process of Microsporogenesis

3.5 Structure of Mature Pollen Grain

- The pollen grains represent the male gametophytes.
- Pollen grains are generally spherical and measuring about **25-50mm** diameter.
- It has two walls namely an **outer exine** and **inner intine**.
- Exine is made up of **sporopollenin** which is one of the most resistant organic materials. It can with stand strong acid and alkali. No enzymes that degrade **sporopollenin** is so far known. Because of this property pollen grains can be preserved as fossil.
- Exine has prominent apertures called **germ pores** where sporopollenin is absent.
- Intine is a thin and continuous layer made up of **cellulose** and **pectin**.
- The cytoplasm is surrounded by plasma membrane.

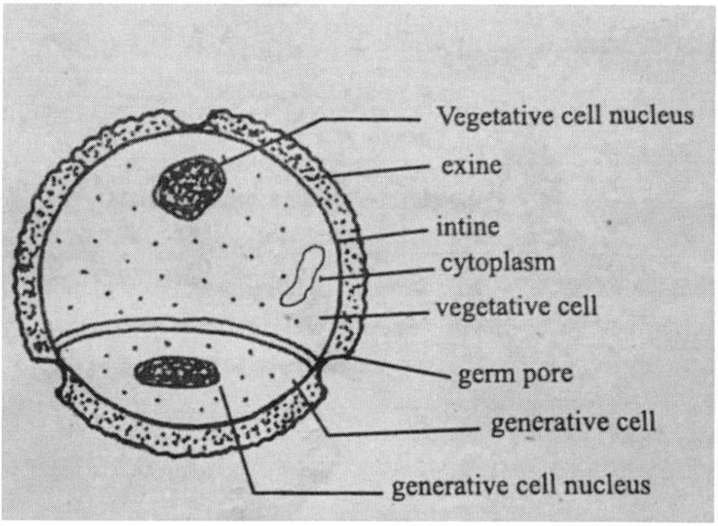

Fig. 3.4: Structure of Pollen Grain

3.6 Pistil or Carpel and its Parts

Pistil or Gynoecium is the female reproductive structure of flower. It consists of **stigma, style and ovary** containing **ovules** inside.

- Each Pistil has 3 parts namely **stigma, style, and ovary.**
- The stigma as a landing platform for **pollen grains.**

- The style is the elongated slender part beneath the stigma. The pollen tube grows through tissue of style.
- The basal bulged part of the pistil is the ovary.
- The ovary is a hallow structure having one or several chambers called ovarian cavity called **locules.**
- After fertilization, the ovary matures into **fruit.**
- Areas where ovules are attached within ovaries are called **placenta.** The arrangement of ovules on the placenta is called **placentation.**
- Arising from the placenta are **megasporangia**, commonly called ovules. The number of ovules in an ovary may be one (wheat, paddy, mango) to many (papaya, water melon, orchids).

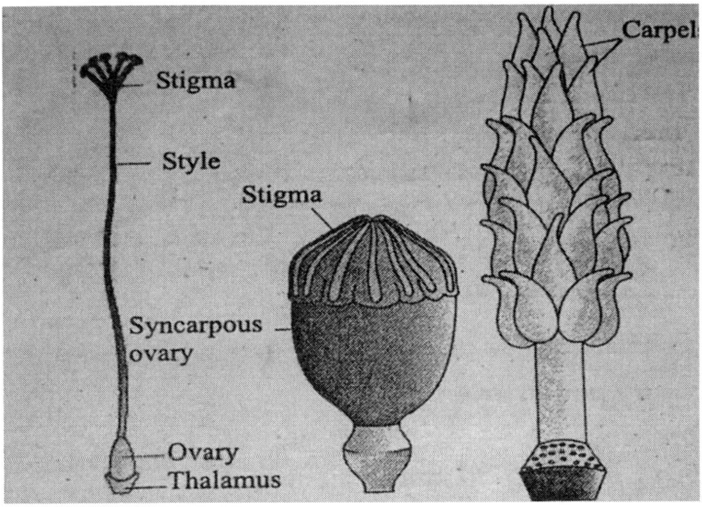

Fig. 3.5: Dissected Flower Showing Pistil

3.6.1. Types of Ovary

a) **Monocarpellary ovary:** A gynoecium with single carpel or pistil. Ex: Pea.

b) **Bicarpellary ovary:** A gynoecium with two carpel or pistil. Ex: *Vinca rosea.*

c) **Tricarpellary ovary:** A gynoecium with three carpel or pistil. Ex: Caster.

d) **Apocarpous ovary:** A gynoecium consists of free carpel (i.e., not fused). Ex: *Michelia champaka.*

e) **Syncarpous ovary:** A gynoecium consisting of fused carpels or pistils. Ex: Lady's finger

3.7 Megasporogenesis

"The process of formation of **megaspores from the megaspore mother cell (MMC)** is called megasporogenesis."

- A diploid **MMC** undergoes meiosis to produce **4 haploid daughter cells.**
- In majority of flowering plants, one of the megaspores is functional while the other three degenerate. Only the functional megaspore develops into the female gametophyte (embryosac). This method of embryo sac formation from a single megaspore is termed as **monosporic development.**
- The nucleus of the functional megaspore divides mitotically to form two nuclei which move to the opposite poles, forming the **2-nucleate embryo sac.**
- Two or more sequential mitotic nuclear divisions result in the formation of **4-nucleate** and later the **8-nucleate** stages of the embryo sac.
- After 8-nucleate stage, cell walls are laid down leading to the organization of the typical female gametophyte or embryosac.
- In the observation, six of the eight nuclei are surrounded by cell walls and organized into cells, the remaining two nuclei called **polar nuclei** are situated at the central region.
- Three cells are grouped together at the **micropylar end** and constitute the **egg apparatus**. The egg apparatus consists of **two synergids** and **one egg cell**. The synergid have cellular thickening at the micropylar tip called **filiform apparatus**, which play an important role in guiding the pollen tubes to penetrate into the synergid.
- Three cells are at the **Chalazal end** are called **antipodals**.
- The large central cell with **two polar nuclei**. Thus, a typical angiosperm embryo sac, at maturity is **8-nucleated and 7-celled.**

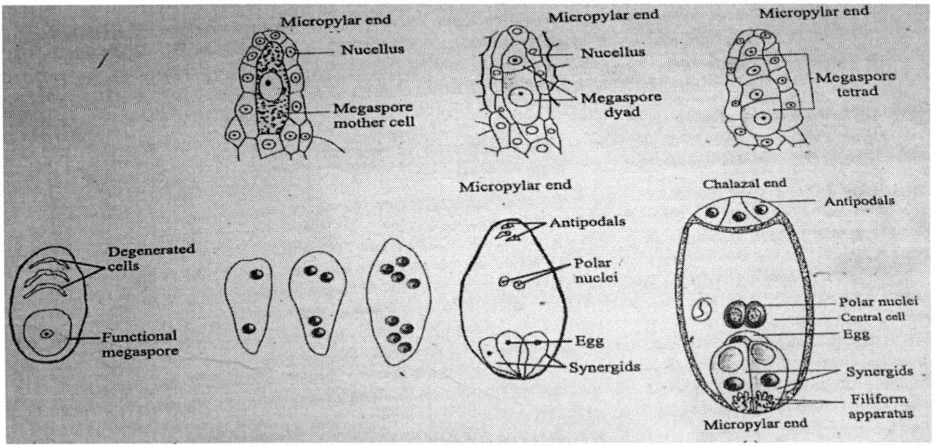

Fig.3.6: Showing Megaspore Mother Cell Dyad, Tetrad Stages, 1, 2, 4 and 8 Nucleated Embryo Sac and Mature Embryo Sac

3.8 Difference between Microsporogenesis and Megasporogenesis

Microsporogenesis	Megasporogenesis
The process of formation of microspores from pollen mother cell through meiosis is called microsporogenesis. Pollen grains are formed at the end of this event. Occurs inside the pollen sac of the anther.	1. The process of formation of megaspores from megaspore mother cells through meiosis is called megaspore -genesis. 2. Egg is, produced inside the embryo sac at the end of this event.

Q2. Arrange the following terms in the correct developmental sequence. Pollen grain, sporogenous tissue, microspore tetrad, pollen mother cell, male gamete.

Answer: Sporogenous tissue → Pollen mother cell → microspore tetrad → pollen grain → male gamete.

3.9 Diagrammetic Representation and Parts of a Typical Angiosperm's Ovule

An ovule is a female mega sporangium where the formation of megaspores takes place.

1. **Funicles**: It is a stalk-like structure which represents the point of attachment of the ovule of the placenta.

2. **Hilum**: It is the point where the body of the ovule is attached to the funicle.
3. **Integuments**: They are the outer layers surrounding the ovule that provide protection to the developing embryo.
4. **Micropyle**: It is a narrow pore formed by the projection of integuments. It makes the point where the pollen tubes enter the ovule at the time of fertilization.
5. **Nucellus**: It is a mass of parenchymatous tissue surrounded by the integuments from the outside. The nucellus provides nutrition to the developing embryo. The embryo sac is located inside the nucellus.
6. **Chalazal**: It is the basal swollen part of the nucellus from where the integuments originate.

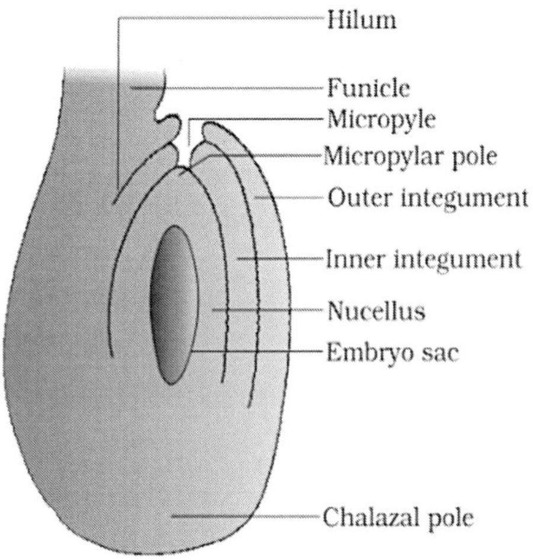

Fig. 3.7: Structure of Typical Ovule

3.10 Types of Ovules

Based on the position of micropyle and funicle, the ovules are of six types they are as fallows.

a) Orthotropus ovule (Atropous ovule):

- It is an straight or erect ovule.
- It is also called Atropous ovule.
- The body of the ovule is upright.

- The funicle, chalazal and micropyle lie on the same vertical line.
- It is most primitive type of ovule.
- Ex: Polygonum.

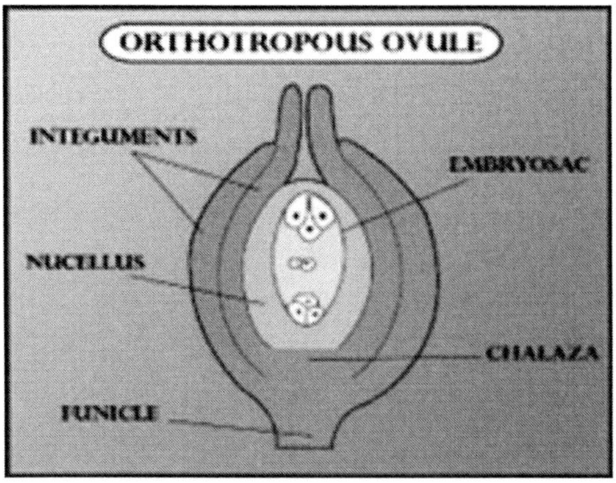

Fig. 3.8: Orthotropus Ovule

b) Anatropous ovule

- It is a most common type of ovule occurs in 82% of angiosperms.
- It is an inverted ovule.
- The micropylar comes nearer to the funicle and Hilum.
- The micropyle and the Hilum face the placenta.
- The chalazal at the opposite end of micropyle.
- Ex: Sunflower, Hibiscus.

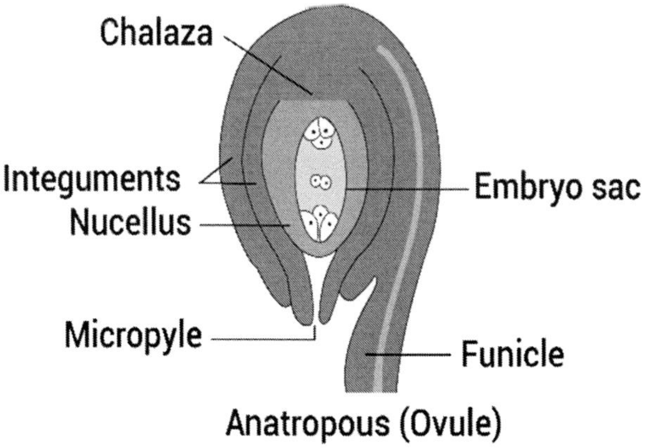

Fig. 3.9: Anatropous Ovule

c) Campylotropous ovule:

- It is curved ovule, the curvature is less than that of anatropous ovule.
- It is a bean shaped ovule.
- The micropyle faces the placenta and it lies near the funicle.
- The micropyle and funicle lie in parallel lines.
- Ex: Brassica, Bean, Green gram.

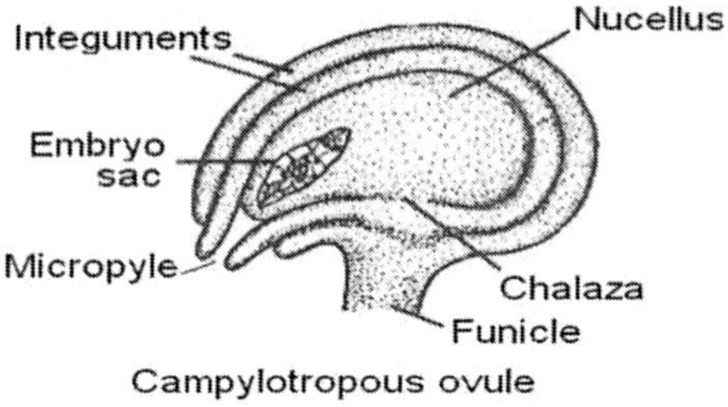

Fig. 3.10: Campylotropous Ovule

d) Hemi-anatropous ovule

- It is a T-shaped ovule.
- The micropyle and chalazal remain at right angles to the funicle.
- Ex: Primula, Ranunculus

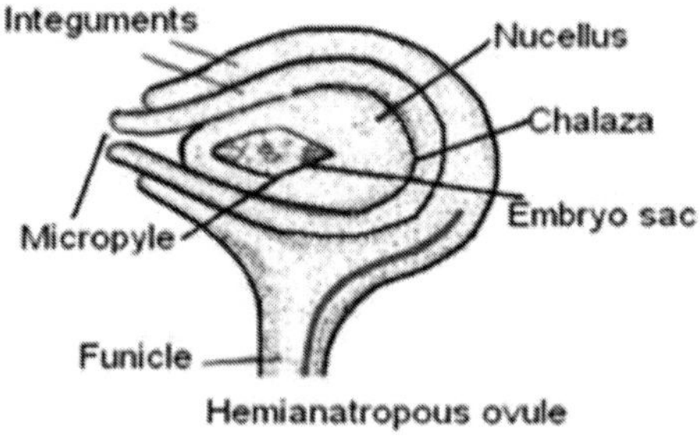

Fig. 3.11: Hemi-anatropous Ovule

e) Amphitropous ovule:

- This type of ovule is looks like a Horse shoe.
- Both ovule and embryo sac curved, hence the name Amphitropous.
- It is intermediate between Orthrotropous and Anatropous ovule.

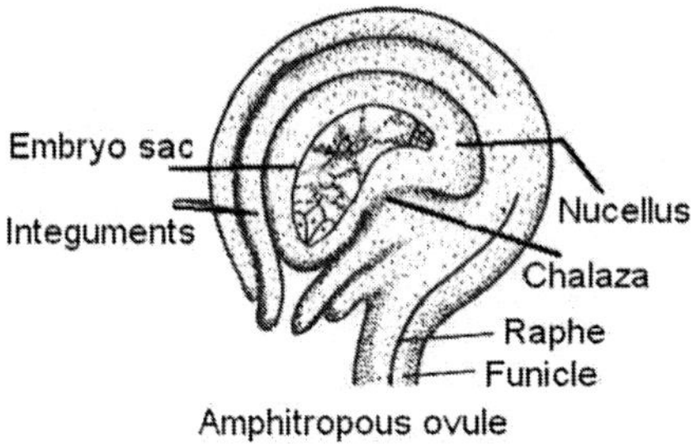

Fig. 3.12: Amphitropous Ovule

f) Circinotropous ovule:

- It is peculiar type of ovule.
- The ovule curves one complete circle, so that the micropyle turns down from upright position and then ovule is straight.
- In the early stage the ovule is straight and then it becomes inverted.
- The growth continuous till the ovule once again becomes straight.
- The funicle is long.

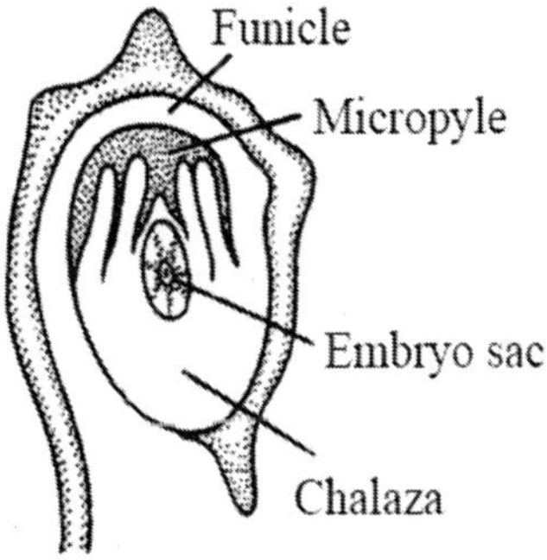

Fig. 3.13: Circinotropous Ovule

3.11 Development of Female Gametophyte (Embryo sac)

Based on the number of megaspores take part in the development of female gametophyte or embryo sac are divided into three types they are

1) Monosporic type
2) Bisporic type
3) Tetrasporic type

1) Monosporic type of development:

- The embryosac derived from one haploid megaspore nucleus is called monosporic type of embryo sac.
- It was first reported by Straus burger in polygonal species. Hence it is also called polygonum type.
- It is a most common type in angiosperms therefore it is called normal.
- In polygonum type the functional megaspore, enlarge in size its nucleus divides mitotically to produce two haploid nuclei.
- The two nuclei are separated by a large central vacuole, the nuclei move towards the opposite pole.
- This is two celled embryosac, each nucleus of pole divides mitotically twice and producing four haploid nuclei at each pole.
- The haploid nucleus undergoes three free nuclear division producing eight haploid nuclei.
- Now organisation of nuclei takes place in the embryo sac.
- The four nuclei present towards the micropyle, three nuclei organise into the egg apparatus. The fourth nucleus act as upper polar nucleus.
- Out of four nuclei present towards the Chalazal end three of them organise into the antipodal cells, the fourth nuclei act as lower polar nucleus.
- The upper and lower polar nuclei move towards the centre of the embryosac and fuse together to form a diploid secondary nucleus or polar nuclei.
- The mature embryo sac or female gametophyte is 8-nucleated and 7-celled. The egg apparatus is consisting of one egg cell and two synergids.

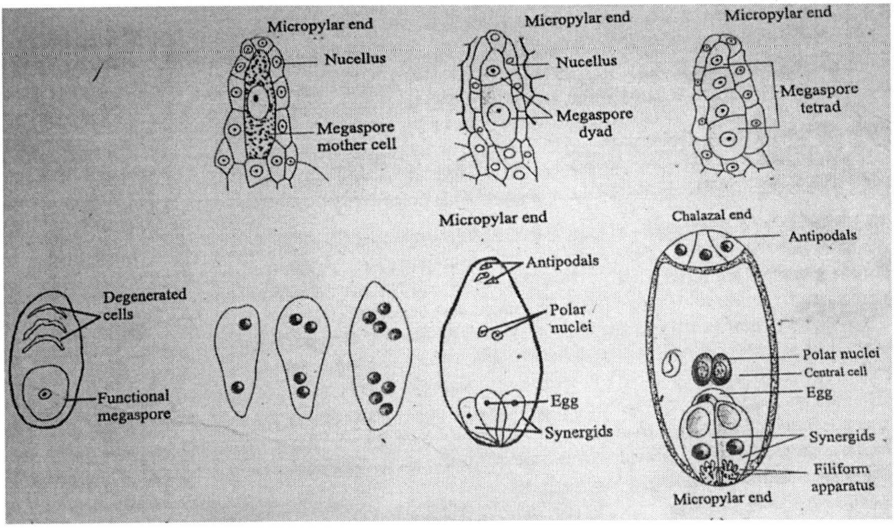

Fig. 3.14: Monosporic Type of Development of Embryosac

2) Bisporic type of development:

- This type of embryosac is derived from two haploid megaspore nuclei (Dyad).
- It was first reported by Straus burger in Allium species (onion), hence it is called Allium type.
- In Allium type the diploid megaspore mother cell divides meiotically and produce two cells.
- The two cells together constitute the Dyad, each cell of the Dyad is equal to a megaspore.
- The upper cell of the dyad becomes degenerates and the lower cell become functional.
- The functional megaspore nucleus divides mitotically produces two nuclei; the two nuclei move towards opposite poles.
- The upper nucleus acts as micropylar nucleus and the lower nucleus is Chalazal nucleus.
- Each nucleus divides twice to produce four nuclei at each pole.
- Now organisation of nuclei takes place in the embryo sac.
- The four nuclei present towards the micropyle, three nuclei organise into the egg apparatus. The fourth nucleus act as upper polar nucleus.

- Out of four nuclei present towards the Chalazal end three of them organise into the antipodal cells, the fourth nuclei act as lower polar nucleus.
- The upper and lower polar nuclei move towards the centre of the embryosac and fuse together to form a diploid secondary nucleus or polar nuclei.
- The mature embryo sac or female gametophyte is 8-nucleated and 7-celled. The egg apparatus is consisting of one egg cell and two synergids.

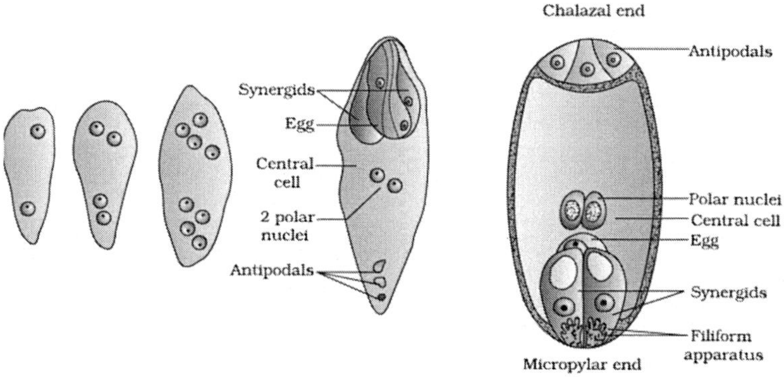

Fig. 3.15: Bisporic type of Development of Embryosac

3) Tetrasporic type of development:

- This type of embryosac derived from four haploid megaspore nuclei. It was reported in Adoxa species hence it is Adoxa type.
- In this embryosac, the diploid nucleus of megaspore mother cell divides meiotically and produces four haploid megaspore nuclei, out of four nuclei two nuclei are arranged at micropylar end and two nuclei at Chalazal end.
- The two nuclei at each pole divide mitotically and produces four nuclei at each pole.
- Now organisation of nuclei takes place in the embryo sac.
- The four nuclei present towards the micropyle, three nuclei organise into the egg apparatus. The fourth nucleus act as upper polar nucleus.
- Out of four nuclei present towards the Chalazal end three of them

organise into the antipodal cells, the fourth nuclei act as lower polar nucleus.

- The upper and lower polar nuclei move towards the centre of the embryosac and fuse together to form a diploid secondary nucleus or polar nuclei.
- The mature embryo sac or female gametophyte is 8-nucleated and 7-celled. The egg apparatus is consisting of one egg cell and two synergids.

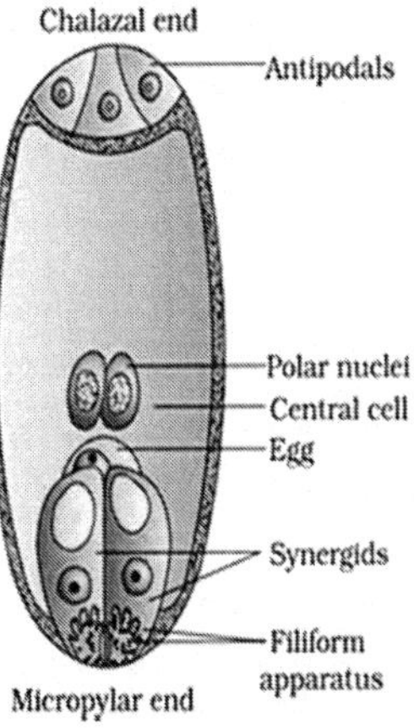

Fig. 3.16: Mature Embryosac

Q3. What are chasmogamous flowers? Can cross-pollination occur in cleistogamous flowers? Give a reason for your answer.

Answer: There are 2 types of flowers present in plants namely oxalis and viola chasmogamous and cleistogamous flowers. Chasmogamous flowers have exposed anthers and stigma like the flowers of other species.

Cross pollination cannot occur in cleistogamous flowers. This is because cleistogamous flowers never open at all. Also, the anther and the stigma lie close to each other in these flowers. Hence only self-pollination is possible in these flowers.

Q4. Mention two strategies evolved to prevent self-pollination in flowers.

Answer: Self-pollination involves the transfer of pollen from the stamen to the pistil of the same flower. Two strategies have evolved to prevent self-pollination in flowers are as follows.

(a) In certain plants, the stigma of the flower has the capability to prevent the germination of pollen grains and hence, prevent the growth of the pollen tube. It is a genetic mechanism to prevent self-pollination called self-incompatibility. This may be between the individuals of the same species or between individuals of different species. Thus, incompatibility prevents breeding.

(b) In some plants, the gynoecium matures before the androecium or vice – versa. This phenomenon is known as protogyny or protandry respectively. This prevents the pollen from coming in contact with the stigma of the same flower.

Q5. What is self-incompatibility? Why does self-pollination not lead to seed formation in self-incompatible species?

Answer: Self-incompatibility is a genetic mechanism in angiosperm that prevents self-pollination. It develops genetic incompatibility between individuals of the same species or between the individuals of different species.

The plants which exhibit this phenomenon have the ability to prevent germination of pollen grains and thus prevents the growth of the pollen tube on the stigma of the flower. This prevents the fusion of the gametes along the development of the embryo. As a result, no seed formation takes place.

Q6. Why do you think the zygote is dormant for some time in a fertilized ovule?

Answer: The zygote is formed by the fusion of the male gamete with the nucleus of the egg cell. The zygote remains dormant for some time and waits for the endosperm to form, which develops from the primary endosperm cell resulting from triple fusion. The endosperm provides food for the developing embryo and after the formation of endosperm, further development of the embryo from the zygote starts.

Q7. Differentiate between

(a) Hypocotyl and epicotyl

(b) Coleoptile and coleorhiza

(c) Integument and testa

(d) Perisperm and pericarp

Answer:

(a) Hypocotyl and epicotyl

Hypocotyl	Epicotyl
1. The portion of the embryonal axis which lies below the cotyledon in a dicot embryo is known as hypocotyl. 2. It terminates with the radicle.	1. The portion of the embryonal axis which lies above the cotyledon in a dicot embryo is known as epicotyl. 2. It terminates with the plumule.

(b) Coleoptile and coleorhiza

Coleoptile	Coleorhiza
It is a conical protective sheath that encloses the plumule in a monocot seed.	It is an undifferentiated sheath that encloses the radicle and the root cap in a monocot seed.

(c) Integument and testa

Integument	Testa
It is the outermost covering of an ovule. It provides protection to it.	It is the outermost covering of a seed.

(d) Perisperm and pericarp

Perisperm	Pericarp
It is the residual nucellus which persists. It is present in some seeds such as beet.	It is the ripened wall of a fruit, which develops from the wall of an ovary.

4

POLLINATION

"Pollination is defined as Transfer of Pollen grains from the anther to the stigma of a pistil or gynoecium is known as pollination".

4.1 Types of Pollination

Depending upon the source of pollen, pollination can be divided into three types.

1. Autogamy (Self-pollination)
2. Geitonogamy
3. Xenogamy (a kind under cross pollination (Allogamy).

1. **Autogamy:** "The transfer of pollen grains from an anther of a flower on to the stigma of same flower" or "Pollination takes place within the same flower."

Conditions for autogamy

- Flower requires synchrony in pollen release and stigma respectively.
- **Chasmogamous flower:** Flowers with exposed anthers and stigma but they should lie close to each other.

  Ex: Viola (Common pansy), Oxalis and Commelina.
- **Cleistogamous flower:** Flowers which do not open at all.

2. **Geitonogamy:** "Transfer of pollen grains from an anther of a flower to the stigma of another flower borne on the same plant.
   - It is functionally cross pollination and requires pollinating agent but genetically they are similar to autogamy since the pollen grains come from the same plant.

3. **Xenogamy**: "Pollination takes place between two flowers borne on two different plants of the same species."
   - **Significance:** It brings genetically different types of pollen grains to the stigma, thus introduces genetic variation.
   - **Example:** Papaya plants exhibit xenogamy because, unisexual flowers borne on two different plants. `

4.2 Agents of Pollination

The agents who achieve cross pollination are abiotic and biotic agents. Abiotic agents are wind and water while biotic agents are animals. Majority of plants use biotic agents for pollination and only a small proportion of plants use abiotic agents. Pollen grains coming in contact with the stigma is a chance factor in both wind and water pollination. To compensate loss of pollen grains, the flowers produce enormous amount of pollen grains than the number of ovules available.

1. **Anemophily:** Pollination by wind is called anemophily and the flower is called anemophilous flower. Pollination by wind is more common amongst abiotic pollination. Wind pollination is very common in grasses.

   Characteristics feature of anemophilous flowers:
   - Wind pollination requires that the pollen grains are light in weight.
   - The flower must produce large number of pollen grains.
   - They should beno sticky and powdery so that they can be transported easily by the the wind currents.
   - They often possess well exposed stamens so that the pollens are easily disappeared into the wind current.
   - The flower must have feathery stigma to easily trap air borne pollen grains.
   - Wind pollinated flowers often have a single ovule in each ovary and numerous flowers packed into inflorescence.
   - A familiar example is corn cob-the tassels we see are the stigma and style which wave in the wind to trap pollen grain.

2. **Hydrophily:** Pollination by water is called Hydrophily and the flower is called Hygrophilous flower.

   Ex: Vallisneria, Hydrilla, Zostera(sea-grasses) etc.

   - Pollination by water is quite rare in flowering plants and is limited to about 30genera and mostly monocotyledons.
   - Water is regular mode of transport for the male gametes among the lower plant group like algae, bryophytes and pteridophytes.
   - In the aquatic plants like water hyacinth and water lily, the flowers emerge above the surface of the water and are pollinated by insects or wind but not by water.

   Note: Mechanism of Pollination in Vallisneria:

   - In Vallisneria, the female flower reach the surface of water by the long stalk and the male flowers or pollen grains released on to the surface of water which are carried by water currents. Some of them eventually on to the female flowers and the stigma finally.
   - Whereas in Sea grasses, female flowers remain submerged in water and the pollen grains are released inside the water. Pollen grains in many such species are longer and ribbon like which are carries by water. Some of them reach the stigma and achieve pollination.
   - In most of the water pollinated species, pollen grains are protected from wetting by a mucilaginous covering.
   - Both wind and water pollinated flowers are not very colourful and not produce nector, hence they are not attracting insects, birds etc.

3. **Zoophily:** Pollination by animals is called Zoophily and the flower is called zoophilous flower.

   Ex: Bees, Butterflies, Flies, Beetles, Ants, Moths, Birds (Sunbirds and Humming birds) and Bats.

   - Among the animals, insects particularly honey bees are the major biotic pollinating agents.
   - Even in larger animals such as lemurs (primates), arboreal (tree dwelling) rodents, reptiles (gecko lizard and garden lizard) have also been reported as pollinators.

- Flowers of animal pollinated plants are specifically adapted for a particular species or animal.
- Majority of insect pollinated flowers are large, colourful, fragrant, and rich in nectar.
- When the flowers are small, the number of flowers is clustered to make them conspicuous.
- The flowers pollinated by flies and beetles secrete foul odours to attract these animals.
- Nectar and pollen grains are the usual rewards to the animals to achieve pollination.
- The body of the animal gets coating with pollen grains, which are generally sticky and comes in contact with the stigma to achieve pollination.
- Other rewards of a flower are providing safe place to lay eggs

  Ex: Tallest flower of Amorphophallus (about 6 feet in height).
- A similar relationship exists between a species of moth and the plant Yucca where both species moth and yucca plant cannot complete their life cycles without each other. The moth deposits its eggs in the locule of the ovary and the flower, in turn, get pollinated by the moth. The larvae of the moth come out the eggs as the seeds start developing.

3.1. Types of Zoophily: Zoophily is further classified into various types

A) Entomophily: Pollination by means of **insects** is called entomophily and the flower is called **entomophilous flower**.

Characteristics features of Entomophilous flower

- Flowers are large, colourful and attractive.
- Flowers are fragrant and produce nectar.
- Pollen grains are sticky in nature.
- When the flowers are small, a number of flowers are clustered into an inflorescence to make them conspicuous.
- The flowers pollinated by flies and beetles secrete fowl odours to attract certain animals.

b) **Ornithophily:** Pollination by means of **birds** is called ornithophily and the flower is called **ornithophilous flower**.

c) **Chiropterophily:** Pollination by means of **bats** is called chiropterophily and the flower is called **Chiropterophilous flower**.

d) **Malacophily:** Pollination by means of **snails** is called malacophily and flower is called **malacophilous flower**.

e) **Myrmecophily:** Pollination by means of **ants** is called myrmecophily and the flower is called **myrmecophilous flower**.

f) **Cantharophily:** Pollination by means of **beetles** is called cantharophily and the flower is called **cantharophilous flower**.

4.3 Pollen Pistil Interaction

- The events from deposition of pollen grains to the stigma until the entry of pollen tube into the ovule are referred to as pollen-pistil interaction.
- Pollination does not guarantee the transfer of the compatible pollen of the same species as the stigma. Often, pollen of the wrong type, either from other species or from the same plant even if it is self-incompatible also deposit on the stigma.
- The pistil has the ability to recognize the pollen of its compatible or incompatible type.
- If it is compatible type, the pistil accepts the pollen and promotes post-pollination events that lead to fertilisation.
- If the pollen is incompatible, the pistil rejects the pollen by preventing pollen germination on the stigma or the growth of pollen tube in the style.
- Acceptance or rejection between pollen grain and pistil is mediated by the chemical components.
- The pollen grains germinated on the stigma to produce a pollen tube through one of the germ pores.
- The contents of the pollen grain move into the pollen tube. Pollen tube grows through the tissues of the stigma and style and reached the ovary.

- In few plant species, the pollen grains are shed at two celled stages (a vegetative cell and a generative cell). In such plants, the generative cell divides and forms the two male gametes during the growth of pollen tube.
- In other plant species where pollen grains shed at 3 celled stages, pollen tube carries the two male gametes.
- Pollen tube, after reaching the ovary, enters the ovule through micropyle and then enters one of the synergids through the filiform apparatus. Recent studies have shown that filiform apparatus present at the micropylar region guides the entry of pollen tube.

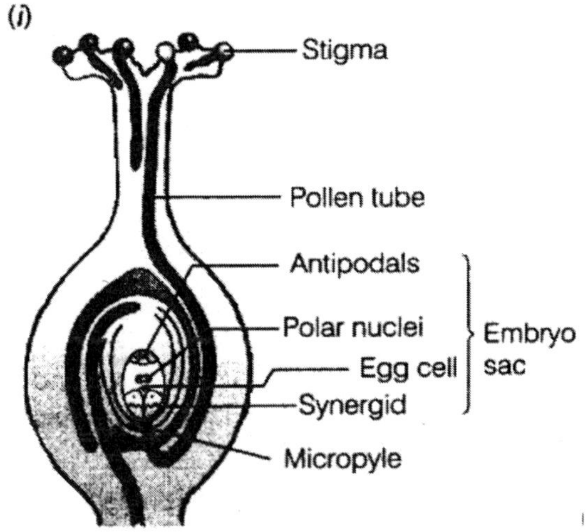

Fig.4.1: Pollen-pistil Interaction

Advantages of Pollen pistil interaction:

- Pollen pistil interaction is a dynamic process involving pollen recognition fallowed by promotion or inhibition of the pollen.
- The knowledge in this area would help plant breeders in manipulating pollen pistil interaction, even in incompatible pollination to get desired hybrids.
- It can be studied easily about the germination of pollen grains by dusting some pollen grains of pea, chickpea, Crotalaria, balsam and Vinca on a glass slide containing a drop of sugar solution (about 10%). After about 15-30 minutes, observe the slide to see pollen tube coming out of the pollen grains.

- It will help the breeders to cross different species and often different genera to combine desirable characters to produce commercially 'superior' varieties.

4.4 Artificial Hybridisation

"Crossing of two genetically different individuals to create new genotype by controlled pollination is called Artificial hybridization."

- It is one of the major approaches of crop improvement programmes.
- In such experiments, only desired pollen grains are used for pollination and the stigma should be protected from contamination of pollen grains. It can be achieved by Emasculation and Bagging technique.

4.5 Emasculation

"Emasculation is the process of removing anthers from bisexual flowers without affecting the female reproductive part (pistil)". Emasculation is performed by plant breeders in bisexual flowers to obtain the desired variety of a plant by crossing a particular plant with the desired pollen grain. After removing the anthers, the flowers are covered with a bag before they open. This ensures that the flower is pollinated by pollen grains obtained from desirable varieties only. Later, the mature, viable and stored pollen grains are dusted on the bagged stigma by breeders to allow artificial pollination to take place and obtain the desired plant variety.

4.5.1 Bagging Technique

Various artificial hybridization technique (various crop improvement programmes involve the removal of the anther from bisexual flowers without affecting the female reproductive part (pistil) through the process of emasculation. Then, these emasculated flowers are wrapped in bags to prevent pollination by unwanted pollen grains. This process is called bagging. This technique is an important part of the plant breeding programme as it ensures that pollen grains of only desired plants are used for fertilization of the stigma to develop the desired plant variably.

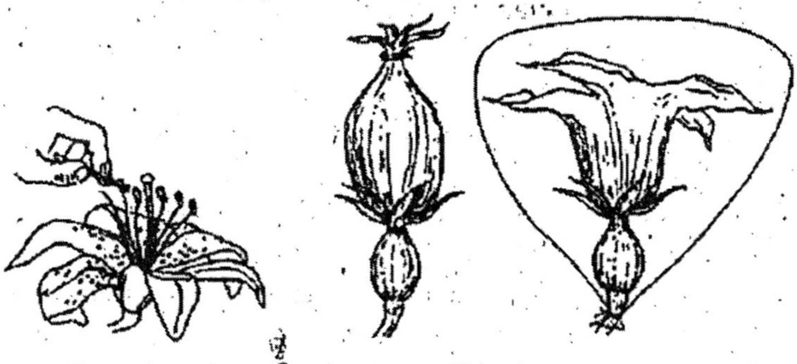

Fig.4.2: Emasculation and Bagging

4.6 Fertilization

"The fusion of haploid male gamete with the haploid female gamete, as a result a diploid cell is formed this process is called Fertilization".

Fertilization occurs inside the female gametophyte of ovule. Fertilization starts with the development of pollen grains on stigma. The fusion of first male gamete with the female gamete produces a diploid zygote which develops into the embryo.

4.6.1 Types of fertilization:

Based on the place of entry of the pollen tube into the ovule, fertilization is of three types namely.

1) **Progamy:** In this type of fertilization the pollen tube enters the ovule through the micropylar end, it is most common type in angiosperms.

   Ex: Bean, Pea, Allium

2) **Chalazogamy:** In this type of fertilization, the pollen tube enters the ovule through chalazal end.

   Ex: Casuarina

3) **Mesogamy:** In this type the pollen tube enters the ovule through the integuments or funicles.

   Ex: Cucurbita

4.7 Triple Fusion or Double Fertilization

"The fusion of two male gamete with two different components of the embryosac namely egg and secondary nucleus is called double fertilization".

Triple fusion is the fusion of the male gamete with 2 polar nuclei inside the embryo sac of the angiosperm. When pollen grains fall on the stigma, they germinate and give rise to the pollen tube that passes through the style and enters the ovule.

- After this, the pollen tube enters between one of the synergids and releases 2 male gametes there.
- Out of these 2 male gametes, one gamete fuses with nucleus of the egg cell and forms zygote (syngamy).
- The other male gametes fuses with 2 polar nuclei in the center form a t**riploid Primary endosperm nucleus (PEN)**.
- This process involves the fusion of 3 haploid nuclei, it is known as triple fusion. It results in the formation of endosperm.
- Since two types of fusions such as syngamy and triple fusion take place in an embryosac, the phenomenon is termed as Double fertilization which is a unique event among flowering plants (Angiosperms).
- The central cell after triple fusion becomes **primary endosperm cell (PEN)** and develop into **endosperm** while zygote develop into an **embryo**.
- The double fertilization involves the fallowing steps.

  a) Germination of pollen grains

  b) Growth of pollen tube

  c) Entry of pollen tube into the ovule

  d) Pollen discharge

  e) Nuclear fission in the embryosac

5

POST-FERTILIZATION EVENTS

After fertilization several changes takes place in the flower, they are called post fertilization changes. These changes take place in complete parts of the flower non-essential changes like weathering of Bracts, Calyx, Carolla but in some cases calyx remains persistent along with the fruits. Stamens either weather or dry one day after the pollination. In Gynoecium the style and stigma get dry up.

Essential changes such as formation of **endosperm** from polar nuclei, **embryo** from fertilized egg, maturation of **ovules into seeds** and **ovary into fruit**, Synergids into X-bodies, Antipodals degenerates.

5.1 Endosperm

"It is a nutritive tissue formed from PEN that nourishes developing embryo".

- The triple cells of an endosperm tissue are filled with reserve food materials and are used for the nutrition of the developing embryo.
- In the most common endosperm development, the PEN undergoes successive nuclear divisions to give rise to free nuclei. The kind of endosperm is called **free nuclear endosperm.**
- Subsequently cell wall formation occurs and the endosperm becomes cellular. This is called **cellular endosperm.**
- The coconut water from tender coconut is nothing but free nuclear endosperm which is made up of thousands of nuclei and the surrounding white kernel is the cellular endosperm.
- Endosperm might either be completely consumed by the developing embryo as in pea, groundnut, beans etc., before seed maturation or it may be persisted in the mature seed as in castor and coconut and it can be used at the time of germination,

5.2 Embryo

"Embryo can be defined as the young sporophytic plant present within a seed (or the minute rudimentary plant contained Within a seed). The structure vin plants that develop from the zygote prior to germination is the **embryo**.

- The early stages of embryo development are called **Embryogeny**.
- Embryo develops at the micropylar end of the embryo sac where the zygote is situated.
- Most of the zygote divide only after the formation of certain amount of endosperm. This is an adaptation to provide assurance for the developing embryo.
- The zygote gives rise to the **proembryo** and subsequently to the **globular, Heart shaped** and **mature embryo**.
- There are us usually two types of embryos i.e. **Dicotyledonous embryo** and **Monocotyledonous embryo**.

5.2.1 Dicotyledonous embryo:

- A typical dicot embryo consists of an embryonal axis and two cotyledons.
- The portion of embryonal axis above the level of cotyledons is the **epicotyl,** which terminates with the **plumule** (stem tip).
- The cylindrical portion below the level of cotyledons is **hypocotyl** that terminates at its lower end in the **radicle** (root tip).
- The root tip is covered with a **root cap.**

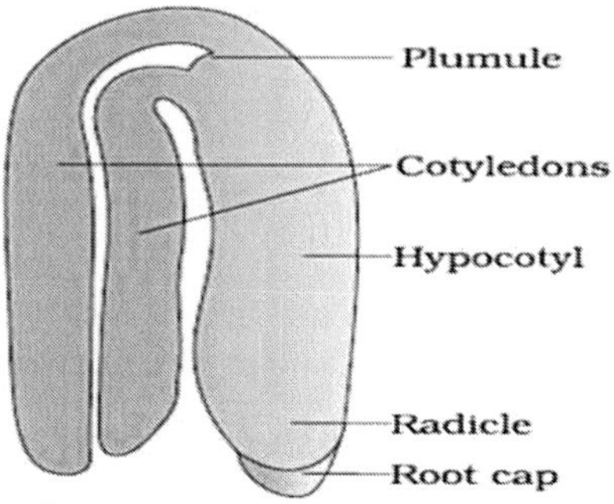

Fig. 5.1: Dicotyledonous Embryo

5.2.2 Different stages of Embryo development in Dicots:

- The zygote situated at the micropylar end of the embryosac divides and develops into an embryo. The early stages of development of embryo are called Embryogeny.
- Most zygote divided only after some amount of endosperm is formed.
- The zygote divided only after some amount of endosperm is formed.
- The zygote undergoes mitotic division to form 2 cells, namely basal cell and apical cell. The basal cell divides further and forms 6-10 celled structure called **suspensor.**
- The terminal cell divides continuously to form a globular embryo which later on becomes heart shaped.

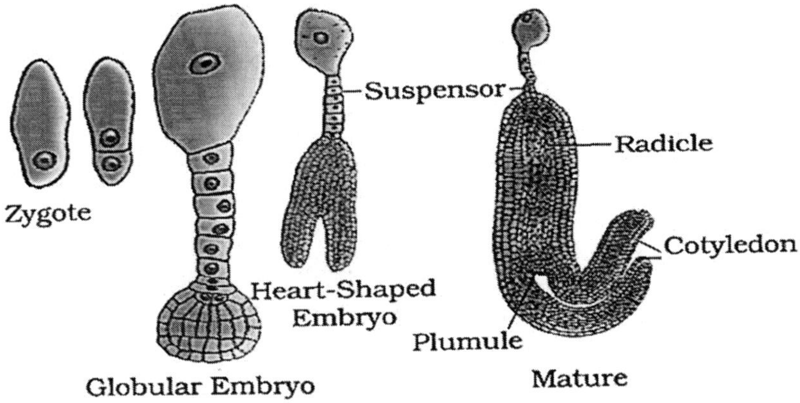

Fig. 5.2: Stages in Embryo Development in a Dicot

5.2.3. Monocotyledonous embryo:

- Monocots possess only one **cotyledon**.
- In the grass family, the cotyledon is called **scutellum** that is situated towards one side of the embryonal axis.
- At its lower end, the embryonal axis above the level of the attachment of scutellum is the **epicotyl.**
- Epicotyl has a shoot apex and few leaves primordial enclosed in a hallow foliar structure called **coleoptile**.

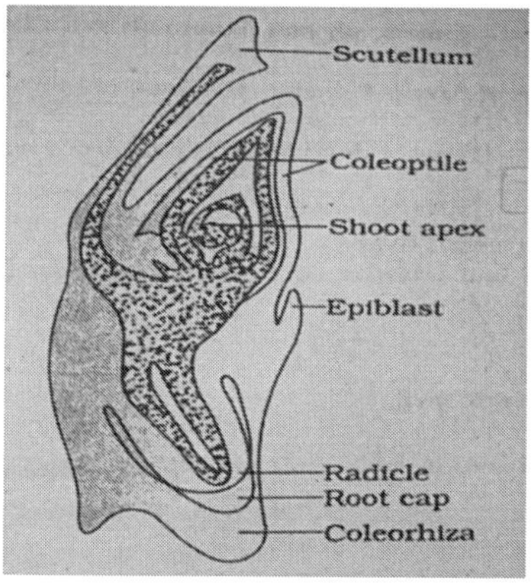

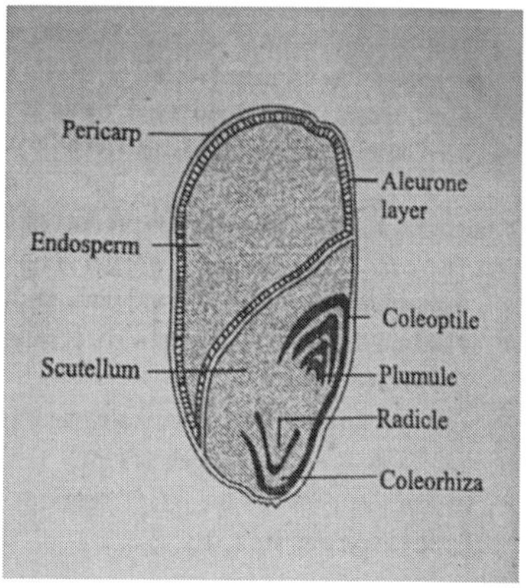

Fig. 5.3: Structure of Monocotyledonous Embryo of Grass and Maize Seed

5.3 Seed

"Seed is a post fertilization product of ovule or seed is a fertilised ovule".

- A seed consists of seed coats, cotyledons and an embryo axis.
- Mature seeds may be **Albuminous or Non-albuminous.**

1) **Albuminous seeds (Endospermic seeds):** The seeds retain a part of endosperm as it is not completely used up during embryogenesis, but remaining endosperm is utilised during seed germination,

Ex: Wheat, Maize, Barley, Castor, Sunflower etc.

2) **Ex-albuminous or non-albuminous seeds (Non-endospermic):** The seeds have no residual endosperm as it is completely consumed during embryogenesis.

Ex: Pea and Ground nut etc.

- Occasionally in some seeds like black pepper and beet, remnants of nucellus are also persistent. This residual persistent nucellus is called **perisperm**.
- Integuments of ovules harden as protective seed coats such as outer **testa** and inner **tegmen.**
- The micropyle remains as a small pore in the seed coat, this facilitates entry of oxygen and water into the seed during germination.

- As the seed matures, its water content is reduced as well metabolic activity and the embryo may enter into a state of inactivity called **Dormancy.**
- When seed gets favourable conditions like adequate moisture, oxygen and suitable temperature, it germinates.

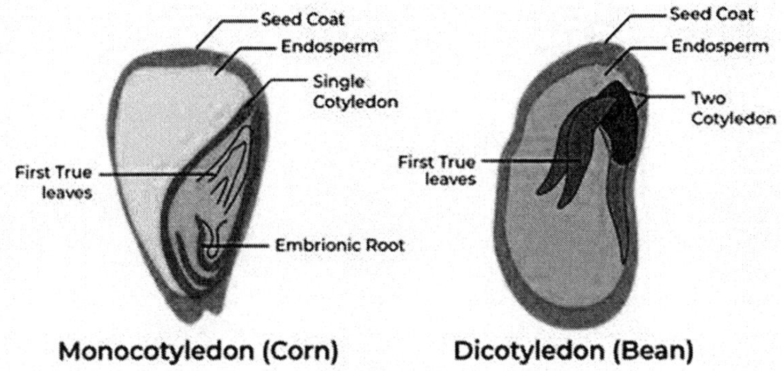

Fig.5.4: Structure of Typical Fruit of Mango and Coconut

5.3.1 Advantages of seeds

- As the seeds have sufficient reserve food, young seedlings are nourished until they are capable of photosynthesis on their own.
- Hard seed coat provides protection to the embryo.
- As they are the products of sexual reproduction, they contribute new genetic combination to new generation that leading to variation which is a necessary tool for evolution.
- The seeds of some plant species remain alive for longar period. Some seeds can remain alive for hundreds of years.
- The oldest seeds that of a lupine, ***Lupinus arcticus*** excavated from Arctic Tundra. The seed germinated flowered after an estimated record of 10,000 years of dormancy.
- A recent record of 2000 years old viable seed is of the date palm ***Phoenix dactylifera*** discovered during archaeological excavation at **King Herod's palace** near **Dead Sea.**

5.4. Fruit, its parts and types

"Fruit is a post fertilization product of ovary. Or "Fruit is a ripened ovary."

- The ovary wall develops into the fruit wall called **pericarp** that consists of three layers such as outer **epicarp,** middle **mesocarp** and inner **endocarp.**
- The fruit that develops from ovary is called **true fruit.**

Ex: Mango, Lemon, Pea etc.

- The fruit that develops from thalamus is called **false fruit or pseudocarp.**

Ex: Cashew, Apple, Strawberry etc.

- In few plant species the ovary develops into fruit without fertilization such fruit is called **Parthenocarpic fruit or seedless fruit.**

Ex: Banana

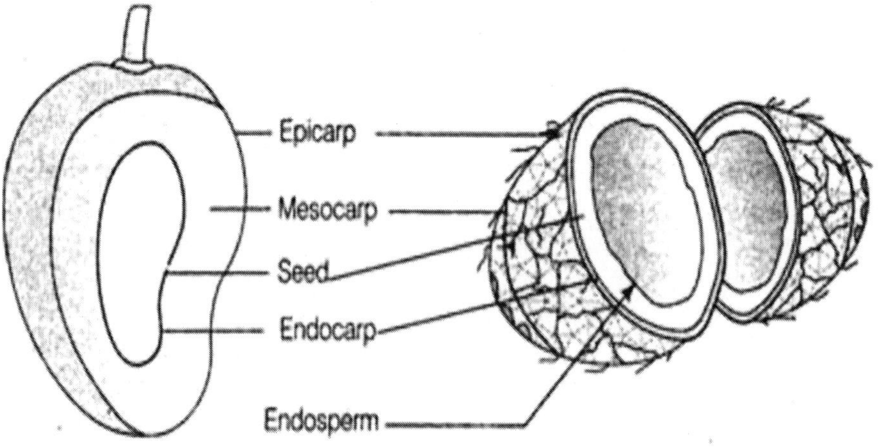

Fig.5.5: Structure of Typical Fruit of Mango and Coconut

5.5. Parthenocarpy

Parthenocarpy is the process of developing fruits without involving the process of fertilization or seed formation. Therefore, the seedless varieties of economically important fruits such as orange, lemon, grapes etc. are produced using this technique. This technique involves inducing fruit formation by the application of plant growth hormone such as auxins. Parthenocarpy is due to the lack of pollination, fertilization and embryo development.

5.5.1. Application of Parthenocarpy:

- Parthenocarpic fruits are generally seedless, so it is applied in horticulture.
- Seedless fruits are superior in size, shape and test.
- Parthenocarpic fruits are highly delicious.
- They have a high demand in the market.

Q12. Why is apple called a false fruit? Which part(s) of the flower forms fruit?

Answer: Fruits derived not from the ovary but from other accessory floral parts are called false fruits. On the contrary, true fruits are those fruits in which fleshy part develop from ovary, but do not consist of the thalamus or any other floral parts. In an apple the fleshy receptacle forms the main edible part. Hence it is a false fruit.

5.6 Apomixis and its importance

Apomixis in the mechanism of seed production without involving the process of meiosis and syngamy. Or Formation of seeds without fertilization is termed as apomixis.

Ex: Species of Asteraceae and Grasses.

- Thus, apomixis is a form of sexual reproduction that **mimic sexual reproduction.**
- There are several ways of development of apomictic seeds.
- In some species, the diploid egg cell is formed without reduction division and develops into embryo without fertilization.
- It plays an important role in hybrid seed production.
- The method of producing hybrid seeds by cultivation is very expensive for farmers. Also, by sowing hybrid seeds, it is difficult to maintain hybrid.
- Characters that segregate during meiosis. Apomixis prevents the loss of specific characters in the hybrid. Also, it is a cost-effective method for producing seeds.
- Then the farmers can keep on using the hybrid seeds to raise new crop year after year and he does not have to buy hybrid seeds every year.

5.6.1 Types of Apomixis

- **Non-recurrent apomixis:** In this the embryos are formed without fertilization from the egg or from any cell of the gametophyte and they are haploid and sterile and they cannot reproduce further.

  Ex: Solanum species (Datura).

- **Recurrent Apomixis:** In this apomixis the embryos are formed without fertilization, they are diploid and fertile, they can reproduce further.

  Ex: Parthenia

- The production of diploid embryo from a cell of the sporophyte without meiosis is called **Apospory.**

- **Sporophytic Budding:** In this type of apomixis embryos arise from the cells of the nucellus or integuments. Here the embryos do not arise from the cells of the gametophyte.

  Ex: Citrus, Opuntia, Mango etc.

- **Changes of flower into vegetative reproduction:** In this case of apomixis the flowers are replaced by bulbils. Bulbils are small Bulblike structure they are the modifications of flowers or vegetative shoot.

5.7 Polyembryony

"Presence of more than one embryo in seed is called polyembryony."

Ex: Citrus and Mango.

As in many citrus and Mango varieties some of the nucellar cells surrounding the embryo sac start dividing, protrude into the embryosac and develop into the embryos. In such species each ovule contains many embryos.

- This phenomenon was initially discovered by Leeuwenhoek (1719).

- Ernst and Schnarf classified the polyembryony into two types.

  1) True polyembryony

  2) False polyembryony

1) **True polyembryony:** In this type, many embryos develop within an embryo sac. It may be cleavage polyembryony, Adventive polyembryony.

In cleavage polyembryony the zygote divides to form an embryogenic mass of cells which splits into two or more embryos.

Ex: Nymphea

In adventive polyembryony, the embryos are formed from the nucellus or integuments.

Ex: Citrus, Mango, Opuntia

The additional embryo is formed from the cells of the embryosac other than the egg.

Ex: Arstolochia

2) **False polyembryony:** In this type, many embryosac fused together to produce many embryos, it is due to the presence of multiple embryo sac within the ovule. As a result, many embryos formed in one seed.

Ex: Lemon, Onion.

6

QUESTIONS AND ANSWERS

Question 1. Name the protective substance present on the pollen envelop to tide over adverse condition.
Answer: Sporopollenin.

Question 2. Name some water pollinated plats.
Answer:

- Fresh water – Vallisneria
- Hydrilla Aquatic – Zostera

Question 3. What technical term is applied to fruits formed without fertilization?
Answer: Parthenocarpy

Question 4. What is a false fruit? Give an example.
Answer: A fruit which is formed by any floral parts of the flower other than ovary, e.g. Apple, pear, cashew nut, etc.

Question 5. List the general characteristics of the pollen grains of wind pollinated plants.
Answer: The pollen grains are smooth, dry, light, non-sticky and may be winged.

Question 6. Which nuclei fuse to give endosperm?
Answer: Polar nuclei.

Question 7. Name the stage of the occurrence of more than one embryo in a seed.
Answer: Polyembryony.

Question 8. The term for the early stages of embryo development.
Answer: Embryogeny

Question 9. Define Microsporogenesis?

Answer: The process of formation of pollen grains or microspore from pollenmothercell(PMC)throughmeiosisiscalledMicrosporogenesis.

Question 10. Define Megasporogenesis?

Answer: The process of formation of **megaspores from the megaspore mother cell (MMC)** is called megasporogenesis.

Question 11. Define Geitonogamy?

Answer: Transfer of pollen grains from an anther of a flower to the stigma of another flower borne on the same plant.

Question 12. Define Xenogamy?

Answer: Pollination takes place between two flowers borne on two different plants of the same species.

Question 13. Define Autogamy?

Answer: The transfer of pollen grains from an anther of a flower on to the stigma of same flower or Pollination takes place within the same flower.

Question 14. Define Cleistogamy?

Answer: It is type of automatic self-pollination of certain plants that can propagate by using non-opening, self-pollinating flowers especially well known in peanuts, peas.

Question 15. Define Chasmogamy?

Answer: It is a reproductive mechanism in plants that helps in the cross pollination of flowers. The flower which shows chasmogamy are known as chasmogamous flowers. The word chasmogamous has been taken from the Greek word which means 'open marriage'.

Question 16. Define Syngamy?

Answer: The fusion of haploid male gamete with the haploid female gamete, as a result a diploid cell is formed this process is known as syngamy.

Question 17. Define Triple fusion or Double fertilization?

Answer: The fusion of two male gamete with two different components of the embryosac namely egg and secondary nucleus is called double fertilization

Question 18. Define Emasculation?

Answer: Emasculation is the process of removing anthers from bisexual flowers without affecting the female reproductive part (pistil).

Question 19. Define nucellus?

Answer: It is nutritive tissue in the ovule that contains the embryo sac, it is made of parenchyma cells.

Question 20. Name the tissue that nourishes pollen mother cells?

Answer: Tapetum.

Question 21. What is cryopreservation?

Answer: It is method of keeping the living cells, tissues and other biological samples in a deep freeze at subzero temperature for the storage or preservation.

Question 22. What is monosporic embryo sac development?

Answer: The embryosac derived from one haploid megaspore nucleus is called monosporic type of embryo sac.

Question 23. Geitonogamy is genetically similar to autogamy. Justify

Answer: The pollen grains are not transferred to a different plant. Hence, due to this there is no genetic variation in the offspring, and identical to parent plant. Hence geitonogamy can be called autogamy as the pollen grains come from the same plant and is genetically similar autogamy.

Question 24. State the function of a suspensor.

Answer: Pushes the developing embryo into the endosperm for absorption of nutrients.

Question 25. Define parthenocarpy.

Answer: These are seedless fruits which are formed without pollination and fertilization.

Question 26. What is fertilization?

Answer: It is the fusion of one male gamete with egg cell and second gamete with polar nuclei in angiosperm

Question 27. State the difference between the endosperm of gymnosperms and angiosperms.

Answer: Endosperm of gymnosperms is haploid gametophyte but in angiosperms it is triploid as it is formed after double fertilization.

Question 28. What is epicotyl?

Answer: Portion of an embryonic axis between the plumule and cotyledon.

Question 29. Define the term geitonogamy.

Answer: A condition where pollen from one flower deposited on the stigma of another flower borne on the same plant.

Question 30. What is the fate of secondary nucleus after fertilization.

Answer: It forms the endosperms.

Question 31. What is allogamy?

Answer: It is the transfer of pollen from one flower to the stigma of another flower on a separate plant of the same species.

Question 32. What develops into a microspore mother cell in a flower?

Answer: Sporogenous cells.

Question 33. What is coleorhiza?

Answer: It is the protective cap over the radicle of maize.

Question 34. What is scutellum?

Answer: It is the single cotyledon of maize.

Question 35. What is funiculus?

Answer: Stalk of the ovule by which it is attached to the placenta.

6.1. Two Marks Questions

Question 1. Differentiate between parthenocarpy and parthenogenesis.

Answer: Parthenocarpy is the formation of fruit without fertilization whereas parthenogenesis is the formation of embryo from unfertilized egg.

Question 2. Draw a labelled diagram of mature pollen grain.

Answer:

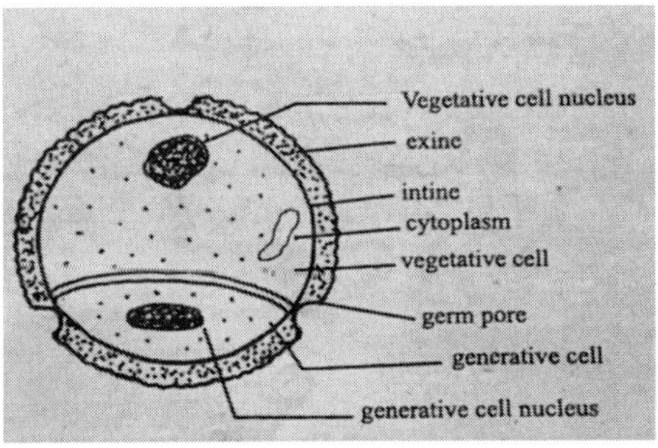

Fig. 6.1: Structure of Pollen Grain

Question 3. What is seed dormancy. Give any 2 advantages.

Answer: Seed dormancy is the condition of the seed when it fails to germinate even though the environmental conditions are favourable for active growth.

Two advantages of dormancy:

- It helps the seed to disseminate in time and space in order to achieve maximum cooperative environment for the survival of species.
- To ensure the successful seed germination under most favorable condition.

Question 4. What is Tapetum? Mention its significance.

Answer: It is innermost layer of the anther wall and is developed from primary parietal cells. It is present below the middle layer. The cells of tapetum are large, radially elongated multinucleated and contain dense cytoplasm.

Significance of Tapetum:

- It provides nutrition for the developing pollen grains.
- It plays an important role in the pollen wall formation.
- The Ubisch bodies present in the tapetal cell contribute a material called sporopollenin, which is involved in the formation Exine of pollen grains.

Question 5. Mention any two biotic and abiotic agents that effect pollination in plants.

Answer:

a) Biotic agents: Pollination takes place by animals are called as biotic agents. It includes birds, insects, flies.

b) Abiotic agents: Pollinators such as wind, water, rain etc.

Question 6. What is cleistogamous flower? Give an example.

Answer: Flowers which do not open at all Ex: Viola (common pansy), Oxalis and Commelina.

Question 7. Define chasmogamous flower with suitable example.

Answer: Flower with exposed anthers and stigma but they should lie close to each other. Ex: Viola (common pansy), Oxalis and Commelina.

Question 8. Define perisperm with an example?

Answer: Nucellus remnants can also be found in some seeds. The perisperm is the nucellus that remains after fertilization. Ex: Black pepper and beet.

Question 9. What is Polyembryony? Give an example.

Answer: Presence of more than one embryo in seed is called polyembryony.

Ex: Citrus and Mango.

Questions 10. What are false fruits? Give example.

Answer: The fruits which develop from parts other than ripened ovary are called false fruits. Eg: fruits of apple and pear develop from fleshy thalamus.

Question 11. Write the significance of double fertilization.

Answer: Double fertilization leads to the development of triploid endosperm which provides nourishment to the developing healthy seed and this triploid endosperm compensates for the extreme reduction of female gametophyte in angiosperms.

Question 12. What is filiform apparatus? What is their function.

Answer: It is finger like projection which guides the entry of pollen tube and helps in the release of pollen grains, and also help in absorption and conduction of food materials from nucellus to embryo sac.

Question 13. Name the two post fertilization events in angiosperms.

Answer: Non-essential changes like weathering of Bracts, Calyx, Carolla but in some cases calyx remains persistent along with the fruits. Stamens either weather or dry one day after the pollination. In Gynoecium the style and stigma get dry up.

Essential changes such as formation of **endosperm** from polar nuclei, **embryo** from fertilized egg, maturation of **ovules into seeds** and **ovary into fruit**, Synergids into X-bodies, Antipodals degenerates.

Question 14. Differentiate between non-albuminous and albuminous seeds with examples.

Answer:

- Non albuminous seeds have no residual endosperm as it is completely consumed during embryo development. Eg: Pea, ground nut.

- Albuminous seeds retain a part of endosperm as it is not completely used up during embryo development. Eg: Wheat, maize, barley, castor.

Question 15. Write the difference between coleoptile and coleorrhiza.

Answer:

- In the monocot seed, the region of embryonic axis below the cotyledon is the radicle covered with protective sheath is coleorrhiza.
- Above the point of attachment of the cotyledon, the embryonic axis becomes the plumule which is enclosed by a leaf like covering called coleoptile.

6.2.Three Marks Question

Question 1. Write four characteristics of wind pollinated flowers give two examples.

Answer: Pollination by wind is called anemophily and the flower is called anemophilous flower. Pollination by wind is more common amongst abiotic pollination. Wind pollination is very common in grasses.

Characteristics feature of anemophilous flowers:

- Wind pollination requires that the pollen grains are light in weight.
- The flower must produce large number of pollen grains.
- They should beno sticky and powdery so that they can be transported easily by the the wind currents.
- They often possess well exposed stamens so that the pollens are easily disappeared into the wind current.

Question 2. Write four characteristics of insect pollinated flowers give two examples.

Answer: Pollination by means of i**nsects** is called entomophily and the flower is called **entomophilous flower**.

Characteristics features of Entomophilous flower

- Flowers are large, colourful and attractive.
- Flowers are fragrant and produce nectar.
- Pollen grains are sticky in nature.

- When the flowers are small, a number of flowers are clustered into an inflorescence to make them conspicuous.

Question 3. What are out breeding devices? Explain briefly the various outbreeding devices in plants.

Answer: Continued self-pollination results in inbreeding depression. In order to prevent this, flowering plants have developed many devices to discourage self-pollination and to encourage cross pollination.

- In some species, pollen release and stigma receptivity are not synchronized. Either the pollen is released before the stigma becomes receptive or stigma receptive before the release of pollen.
- In some other species anther and stigma are placed at different position (Heterostyly), so that the pollen cannot come in contact with the stigma of the same flower.
- Self-compatibility is genetic mechanism and to prevent autogamy but not geitonogamy.
- Unisexuality is another device to prevent self-pollination. Ex: Castor and maize are monoecious where unisexual male and female flowers borne on same plant prevents autogamy but not geitonogamy.
- In several species like papaya, male and female flowers present on different plants (diecious). This condition prevents both autogamy and geitonogamy
- Dichogamy, in bisexual flower male and female reproductive parts mature at different times.
- Protandry, Androecium matures earlier to gynoecium. Ex: Cotton and sunflower.
- Protogyny, Gynoecium matures earlier than androecium. Ex: 4 'O' clock plant.
- Herkogamy, there is physical barrier between anther and stigma can prevents self-pollination. Ex: Calotropis.

Question 4. Write diagrammatic representation of embryogenesis in a dicot.

Answer: Dicotyledonous embryo:

- A typical dicot embryo consists of an embryonal axis and two cotyledons.
- The portion of embryonal axis above the level of cotyledons is the **epicotyl,** which terminates with the **plumule** (stem tip).
- The cylindrical portion below the level of cotyledons is **hypocotyl** that terminates at its lower end in the **radicle** (root tip).
- The root tip is covered with a **root cap.**

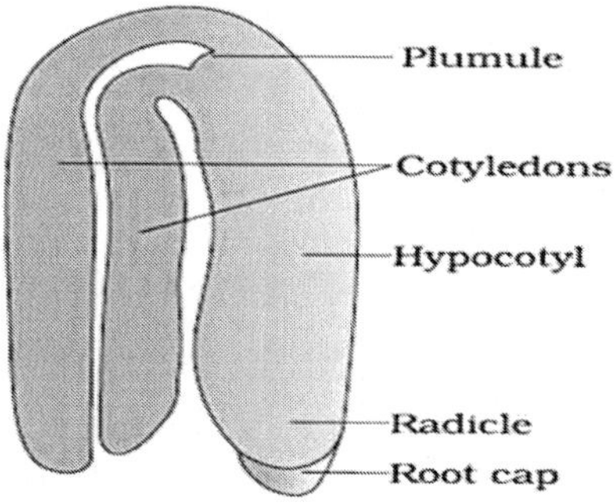

Fig. 6.2: Dicotyledonous Embryo

Question 5. What is endosperm? Distinguish between free nuclear and cellular endosperm with an example for each.

Answer: Endosperm: "It is a nutritive tissue formed from PEN that nourishes developing embryo".

- The triple cells of an endosperm tissue are filled with reserve food materials and are used for the nutrition of the developing embryo.
- In the most common endosperm development, the PEN undergoes successive nuclear divisions to give rise to free nuclei. The kind of endosperm is called **free nuclear endosperm.**
- Subsequently cell wall formation occurs and the endosperm becomes cellular. This is called **cellular endosperm.**
- The coconut water from tender coconut is nothing but free nuclear endosperm which is made up of thousands of nuclei and the surrounding white kernel is the cellular endosperm.

- Endosperm mat either be completely consumed by the developing embryo as in pea, groundnut, beans etc., before seed maturation or it may be persisted in the mature seed as in castor and coconut and it can be used at the time of germination,

Question 6. What is Apomixis? Give example and mention its significance in agriculture.

Answer: Apomixis in the mechanism of seed production without involving the process of meiosis and syngamy. Or Formation of seeds without fertilization is termed as apomixis. **Ex:** Species of Asteraceae and Grasses.

Significance:

- Thus, apomixis is a form of sexual reproduction that **mimic sexual reproduction.**
- There are several ways of development of apomictic seeds.
- In some species, the diploid egg cell is formed without reduction division and develops into embryo without fertilization.
- It plays an important role in hybrid seed production.
- The method of producing hybrid seeds by cultivation is very expensive for farmers. Also, by sowing hybrid seeds, it is difficult to maintain hybrid.

6.3. Five Marks Questions

Question 1. Describe the process of development of dicotyledonous embryo.

Answer: The process of development of mature embryo from diploid zygote is called embryogenesis. After fertilization the zygote of the ovule divides transversely into two cells – a small apical cell and a large basal cell. The basal cell lying towards the Micropyle divides in one direction into a row of cells called a suspensor. It pushes the developing embryo into the endosperm for absorption of nutrients.

The apical cell located towards the antipodal end of the zygote undergoes two vertical divisions and one transverse division to form the embryonal mass. The cells of the embryonal mass divide repeatedly and the various parts of the embryo are formed. The anterior cells of the embryonal mass form _ plumule and two cotyledons. The main part of the radicle and the hypocotyl are formed from the posterior embryonal mass cells.

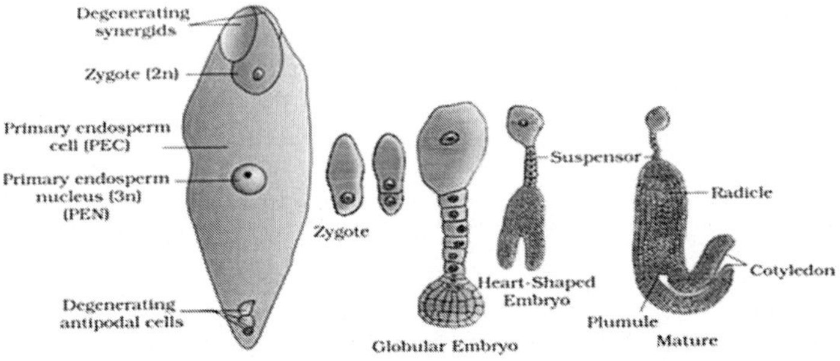

Fig. 6.3: Different Stages of Development of Dicot Embryo

Question 2. Trace the events that would take place in a flower from the time the pollen grain of the same species falls on the stigma up to the completion of fertilization.

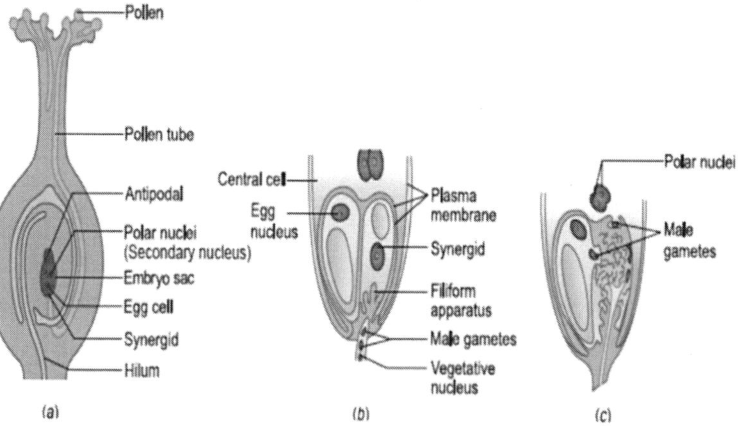

Fig. 6.4:

The pollen grain develops on the stigma stimulated by the secretion of the stigma. The intine grows out as a protuberance through one of the germ pores. This out growth continues to grow as pollen tube. The nucleus of the vegetative cell moves into it followed by the generative cell. The generative cell divides into two male gamete and moves to the tip of the pollen tube.

The pollen tube secretes enzyme to digest the tissues of the style and enters the ovule through micropylar end and discharges the two male gametes into one of the synergids of the embryo sac. One of the male gametes fuses with the ovum to form a zygote. This process is called syngamy. The other male gamete fuses with the secondary nucleus to form primary endosperm nucleus. This process is called triple fusion. Since there are two fusions (syngamy and triple fusion) inside the ovule during fertilization, it is known as double fertilization.

Question 3. Draw a labelled diagram of a T.S of a dehisced anther.

Answer:

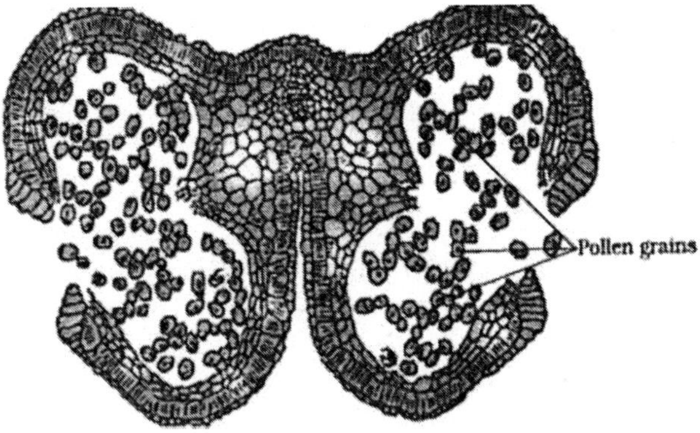

Fig. 6.5: Dehisced Anther

Question 4. Describe the structure of a typical embryo sac found in flowering plant. Why is it generally referred to as monosporic?

Answer: A typical embryo sac is 8 nucleate and 7 celled. Six of the eight nuclei become surrounded by cell wall and are organized into cells. Two of the nuclei, called polar nuclei remain in the center of the large central cell. Three cells are grouped together at the micropylar end forming the egg apparatus. Of these, one is the female gamete and two are synergids; the synergids have special cellular thickening called filiform apparatus at the micropylar tip.

Three cells are grouped at the Chalazal end; they are called antipodal. The large cell in the center of the embryo sac is the central cell. Later 2 polar nuclei in the center cell fused to form a diploid secondary nucleus or endosperm nucleus. Thus, the embryo sac of flowering plants is 8 nucleate 7 celled at maturity. This type of embryo sac is called monosporic because it is formed from only one of the 4 megaspores.

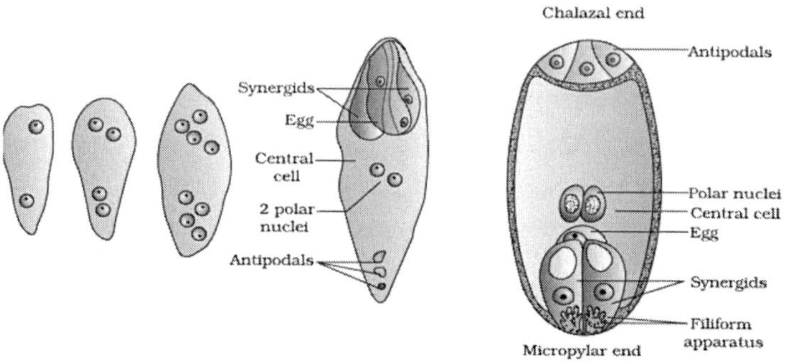

Fig. 6.6: Monosporic Type of Development of Embryo

Question 5. Define pollination. Describe the different stages types of pollination in angiosperms and mention their significance.

Answer: Types of Pollination:

Depending upon the source of pollen, pollination can be divided into three types.

- Autogamy (Self-pollination)
- Geitonogamy
- Xenogamy (a kind under cross pollination (Allogamy).

1. **Autogamy:** "The transfer of pollen grains from an anther of a flower on to the stigma of same flower" or

   "Pollination takes place within the same flower."

   Conditions for autogamy

   - Flower requires synchrony in pollen release and stigma respectively.
   - **Chasmogamous flower:** Flowers with exposed anthers and stigma but they should lie close to each other.

   Ex: Viola (Common pansy), Oxalis and Commelina.

   - **Cleistogamous flower:** Flowers which do not open at all.

2. **Geitonogamy:** "Transfer of pollen grains from an anther of a flower to the stigma of another flower borne on the same plant.

   - It is functionally cross pollination and requires pollinating agent but genetically they are similar to autogamy since the pollen grains come from the same plant.

3. **Xenogamy**: "Pollination takes place between two flowers borne on two different plants of the same species."

   - **Significance:** It brings genetically different types of pollen grains to the stigma, thus introduces genetic variation.
   - **Example:** Papaya plants exhibit xenogamy because, unisexual flowers borne on two different plants. `

Question 6. With neat labelled diagram describe the structure of T.S of Mature anther (Microsporangium).

Answer: Microsporangium:

- In a **T.S** it appears almost circular in outline.
- It is surrounded by four wall layers called **epidermis, endothecium, middle layer** and **tapetum**.
- The outer three wall layers perform the function of protection and help in dehiscence of anther to release pollen grains.
- The innermost layer **tapetum** nourishes the developing pollen grain.
- Cells of tapetum possess dense cytoplasm and have more than one nucleus.
- In young **anther, microsporangium** contains **sporogenous tissue** at the centre which undergoes meiosis to produce **microspore tetrads** which further develop into pollen grains.

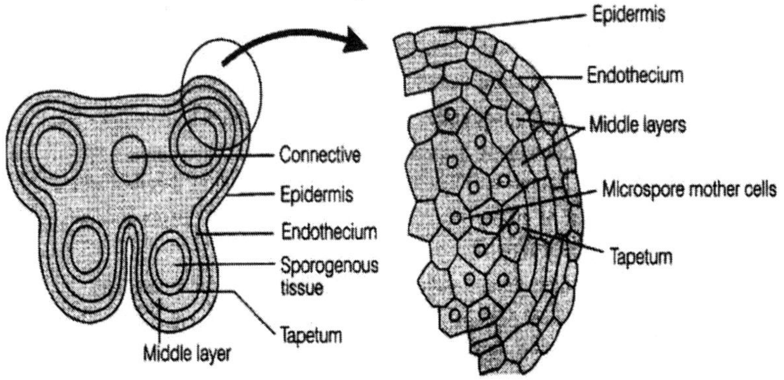

Fig. 6.7: Tis of Mature Anther

Question 7. Write a neat labelled diagram describe the structure of anatropous ovule.

Answer: An ovule is a female mega sporangium where the formation of megaspores takes place.

1. **Funicles:** It is a stalk-like structure which represents the point of attachment of the ovule of the placenta.
2. **Hilum**: It is the point where the body of the ovule is attached to the funicle.
3. **Integuments:** They are the outer layers surrounding the ovule that provide protection to the developing embryo.
4. **Micropyle:** It is a narrow pore formed by the projection of integuments. It makes the point where the pollen tubes enter the ovule at the time of fertilization.
5. **Nucellus:** It is a mass of parenchymatous tissue surrounded by the integuments from the outside. The nucellus provides nutrition to the developing embryo. The embryo sac is located inside the nucellus.
6. **Chalazal:** It is the basal swollen part of the nucellus from where the integuments originate.

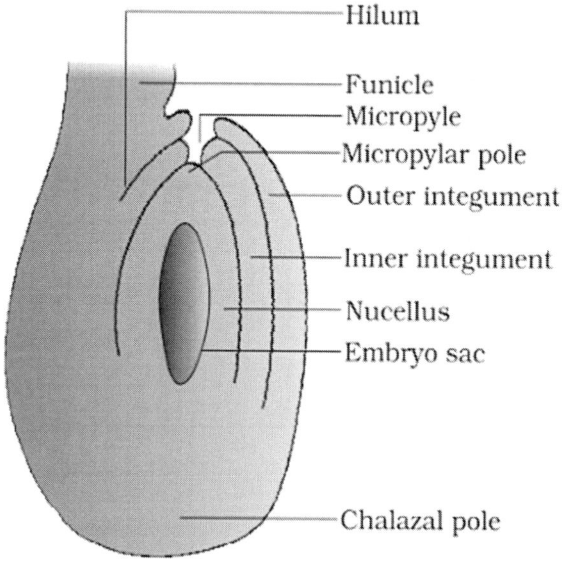

Fig. 6.8: Structure of Anatropous Ovule

Question 8. Describe how does pollination occurs in *Vallisneria* and sea grasses.

Answer: Mechanism of Pollination in Vallisneria:

- In Vallisneria, the female flower reach the surface of water by the long stalk and the male flowers or pollen grains released on to the surface of water which are carried by water currents. Some of them eventually on to the female flowers and the stigma finally.
- Whereas in Sea grasses, female flowers remain submerged in water and the pollen grains are released inside the water. Pollen grains in many such species are longer and ribbon like which are carries by water. Some of them reach the stigma and achieve pollination.
- In most of the water pollinated species, pollen grains are protected from wetting by a mucilaginous covering.
- Both wind and water pollinated flowers are not very colourful and not produce nector, hence they are not attracting insects, birds etc.

Question 9. Describe the phenomenon of double fertilisation in angiosperms.

Answer: Triple fusion or Double fertilization: "The fusion of two male gamete with two different components of the embryosac namely egg and secondary nucleus is called double fertilization".

Triple fusion is the fusion of the male gamete with 2 polar nuclei inside the embryo sac of the angiosperm. When pollen grains fall on the stigma, they germinate and give rise to the pollen tube that passes through the style and enters the ovule.

- After this, the pollen tube enters between one of the synergids and releases 2 male gametes there.
- Out of these 2 male gametes, one gamete fuses with nucleus of the egg cell and forms zygote (syngamy).
- The other male gametes fuses with 2 polar nuclei in the center form a **triploid Primary endosperm nucleus (PEN)**.
- This process involves the fusion of 3 haploid nuclei, it is known as triple fusion. It results in the formation of endosperm.
- Since two types of fusions such as syngamy and triple fusion take place in an embryosac, the phenomenon is termed as Double

fertilization which is a unique event among flowering plants (Angiosperms).

- The central cell after triple fusion becomes **primary endosperm cell (PEN)** and develop into **endosperm** while zygote develop into an **embryo**.
- The double fertilization involves the fallowing steps.

a) Germination of pollen grains

b) Growth of pollen tube

c) Entry of pollen tube into the ovule

d) Pollen discharge

e) Nuclear fission in the embryosac

Question 10. Explain Megasporogenesis with a neat labelled diagram.

Answer: "The process of formation of **megaspores from the megaspore mother cell (MMC)** is called megasporogenesis."

- A diploid **MMC** undergoes meiosis to produce **4 haploid daughter cells.**
- In majority of flowering plants, one of the megaspores is functional while the other three degenerate. Only the functional megaspore develops into the female gametophyte (embryosac). This method of embryo sac formation from a single megaspore is termed as **monosporic development.**
- The nucleus of the functional megaspore divides mitotically to form two nuclei which move to the opposite poles, forming the **2-nucleate embryo sac.**
- Two more sequential mitotic nuclear divisions result in the formation of **4-nucleate** and later the **8-nucleate** stages of the embryo sac.
- After 8-nucleate stage, cell walls are laid down leading to the organization of the typical female gametophyte or embryosac.
- In the observation, six of the eight nuclei are surrounded by cell walls and organized into cells, the remaining two nuclei called **polar nuclei** are situated at the central region.

- Three cells are grouped together at the **micropylar end** and constitute the **egg apparatus**. The egg apparatus consists of **two synergids** and **one egg cell**. The synergid have cellular thickening at the micropylar tip called **filiform apparatus**, which play an important role in guiding the pollen tubes to penetrate into the synergid.
- Three cells are at the **Chalazal end** are called **antipodals**.
- The large central cell with **two polar nuclei**. Thus, a typical angiosperm embryo sac, at maturity is **8-nucleated and 7-celled.**

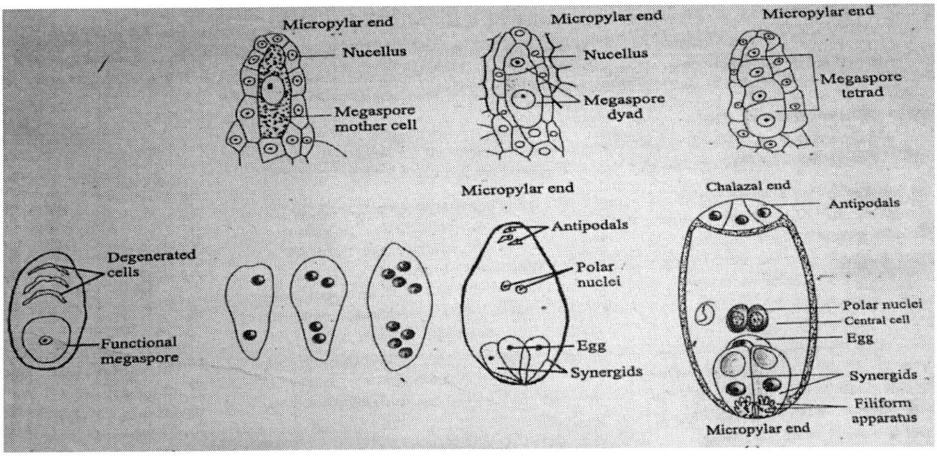

Fig. 6.9: Structure of Anatropous Ovule Showing megaspore mother cell dyad, tetrad stages, 1, 2, 4 and 8 nucleated embryo sac and mature embryo sac

Question 11. Write a note on different types of ovules?

Answer: Based on the position of micropyle and funicle, the ovules are of six types they are as fallows.

a) **Orthotropus ovule (Atropous ovule):**

- It is an straight or erect ovule.
- It is also called Atropous ovule.
- The body of the ovule is upright.
- The funicle, chalazal and micropyle lie on the same vertical line.
- It is most primitive type of ovule.
- Ex: Polygonum.

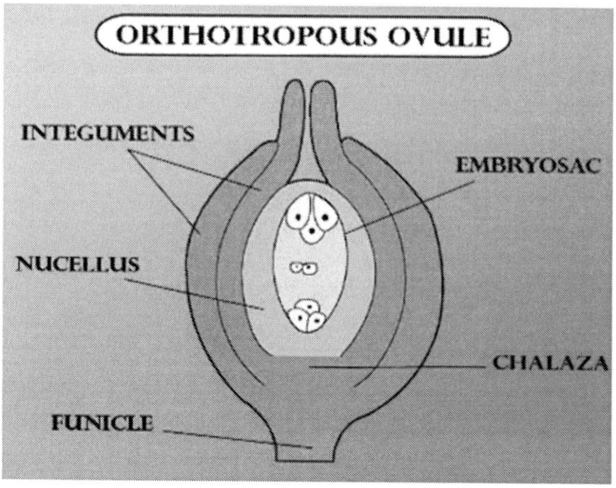

Fig. 6.10: Orthotropus Ovule

b) Anatropous ovule:

- It is a most common type of ovule occurs in 82% of angiosperms.
- It is an inverted ovule.
- The micropylar comes nearer to the funicle and Hilum.
- The micropyle and the Hilum face the placenta.
- The chalazal at the opposite end of micropyle.
- Ex: Sunflower, Hibiscus.

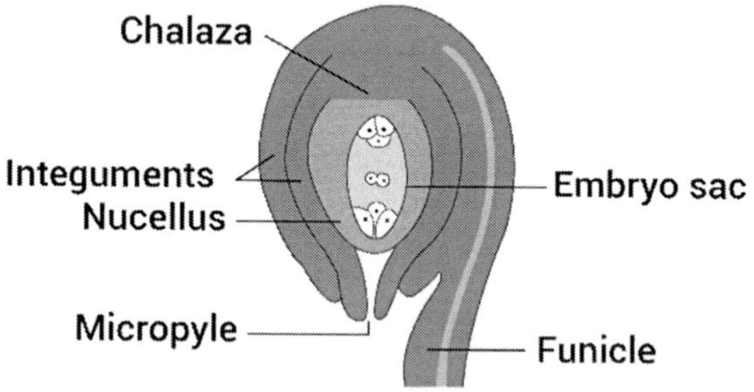

Fig. 6.11: Anatropous Ovule

c) Campylotropous ovule:

- It is curved ovule, the curvature is less than that of anatropous ovule.
- It is a bean shaped ovule.
- The micropyle faces the placenta and it lies near the funicle.

- The micropyle and funicle lie in parallel lines.
- Ex: Brassica, Bean, Green gram.

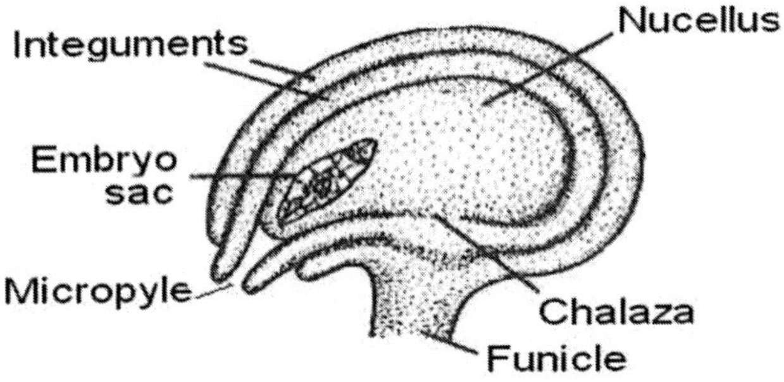

Fig. 6.12: Campylotropous Ovule

d) Hemi-anatropous ovule:

- It is a T-shaped ovule.
- The micropyle and chalazal remain at right angles to the funicle.
- Ex: Primula, Ranunculus

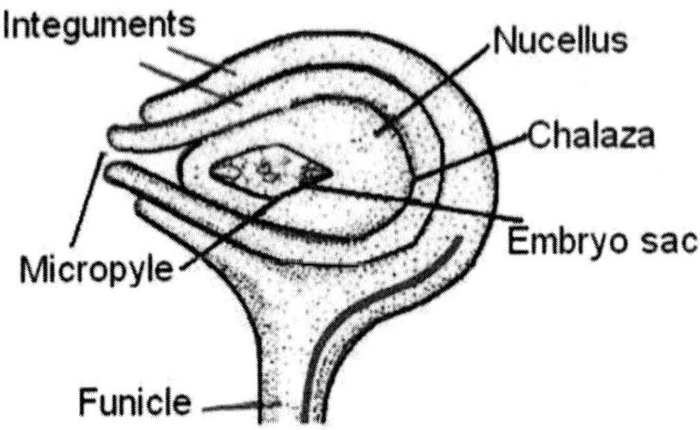

Fig. 6.13: Hemi-anatropous Ovule

e) Amphitropous ovule:

- This type of ovule is looks like a Horse shoe.
- Both ovule and embryo sac curved, hence the name Amphitropous.
- It is intermediate between Orthrotropous and Anatropous ovule.

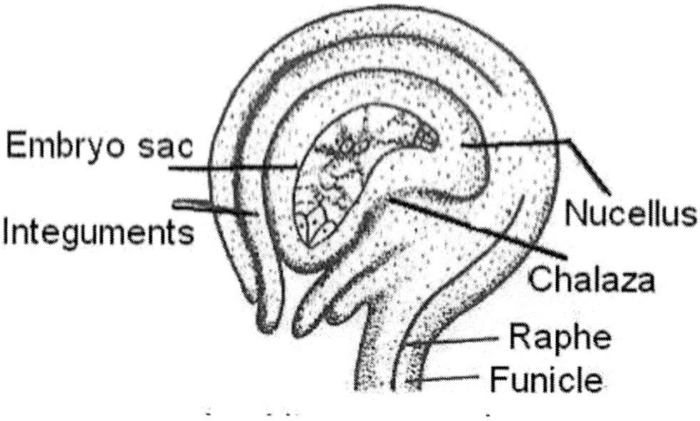

Fig. 6.14: Amphitropous Ovule

f) Circinotropous ovule:

- It is peculiar type of ovule.
- The ovule curves one complete circle, so that the micropyle turns down from upright position and then ovule is straight.
- In the early stage the ovule is straight and then it becomes inverted.
- The growth continuous till the ovule once again becomes straight.
- The funicle is long.

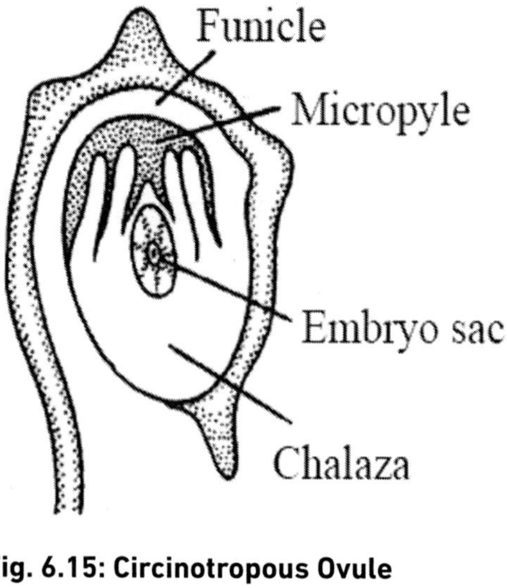

Fig. 6.15: Circinotropous Ovule

7

MULTIPLE CHOICE QUESTIONS

Question 1. Which of the following groups is formed only of the hermaphrodite organisms?

(a) Earthworm, tapeworm, housefly, frog

(b) Earthworm, tapeworm, sea horse, housefly

(c) Earthworm, leech, sponge, roundworm

(d) Earthworm, tapeworm, leech, sponge

Answer: (d) Earthworm, tapeworm, leech, sponge

Question 2. Which of the following options shows bisexual animals only?

(a) Amoeba, sponge, leech

(b) Sponge, cockroach, Amoeba

(c) Earthworm, sponge, leech

(d) Tapeworm, earthworm, honeybee

Answer: (c) Earthworm, sponge, leech

Question 3. Read the following statements and select the incorrect one.

(a) Cucurbits and coconuts are monoecious plants.

(b) Papayas and date palms are dioecious plants.

(c) Leeches and tapeworms are bisexual animals.

(d) Sponges and coelenterates are unisexual animals.

Answer: (d) Sponges and coelenterates are unisexual animals.

Question 4. Meiosis does not occur in

(a) asexually reproducing diploid individuals

(b) sexually reproducing haploid individuals

(c) sexually reproducing diploid individuals

(d) all of these.

Answer: (a) asexually reproducing diploid individuals

Question 5. A diploid parent plant body produces gametes and a haploid parent plant body produces gametes.

(a) diploid, haploid

(b) haploid, diploid

(c) diploid, diploid

(d) haploid, haploid

Answer: (d) haploid, haploid

Question 6. Which of the following organisms has the highest number of chromosomes?

(a) Housefly

(b) Butterfly

(c) Ophioglossum

(d) Onion

Answer: (c) Ophioglossum

Question 7. In maize, a meiocyte has 20 chromosomes. What will be the number of chromosomes in its somatic cell?

(a) 40

(b) 30

(c) 20

(d) 10

Answer: (c) 20

Question 8. If a butterfly has chromosome number 360 in its meiocyte (2n). What will be the chromosome number in its gametes?

(a) 380

(b) 190

(c) 95

(d) 760

Answer: (b) 190

Question 9. In flowering plants, both male and female gametes are non-motile. The method to bring them together for fertilization is

(a) water

(b) air

(c) pollination

(d) apomixis

Answer: (c) pollination

Question 5. A diploid parent plant body produces gametes and a haploid parent plant body produces gametes.

(a) diploid, haploid

(b) haploid, diploid

(c) diploid, diploid

(d) haploid, haploid

Answer: (d) haploid, haploid

Question 6. Which of the following organisms has the highest number of chromosomes?

(a) Housefly

(b) Butterfly

(c) Ophioglossum

(d) Onion

Answer: (c) Ophioglossum

Question 7. In maize, a meiocyte has 20 chromosomes. What will be the number of chromosomes in its somatic cell?

(a) 40

(b) 30

(c) 20

(d) 10

Answer: (c) 20

Question 8. If a butterfly has chromosome number 360 in its meiocyte (2n). What will be the chromosome number in its gametes?

(a) 380

(b) 190

(c) 95

(d) 760

Answer: (b) 190

Question 9. In flowering plants, both male and female gametes are non-motile. The method to bring them together for fertilization is

(a) water

(b) air

(c) pollination

(d) apomixis

Answer: (c) pollination

Question 10. In which of the following plants, sepals do not fall off after fertilization and remain attached to the fruit?

(a) Brinjal

(b) Cucumber

(c) Papaya

(d) Bitter gourd

Answer: (a) Brinjal

Question 11. Which of the labelled parts in the transverse section of tomato fruit, is/are diploid?

(a) X

(b) Y

(c) Both X and Y

(d) None of these

Answer: (c) Both X and Y

Question 12. The wall of the ovary forms

(a) pericarp

(b) fruit wall

(c) fruit

(d) both (a) and (b).

Answer: (d) both (a) and (b).

Question 13. The term 'clone' cannot be applied to offspring formed by sexual reproduction because

(a) offspring do not possess exact copies of parental DNA

(b) DNA of only one parent is copied and passed on to the offspring

(c) offspring are formed at different times

(d) DNA of parent and offspring are completely different

Answer: (a) offspring do not possess exact copies of parental DNA.

Question 14. The male gametes of rice plant have 12 chromosomes in their nucleus. The chromosome number in the female gamete, zygote and the cells of the seedling will be, respectively

(a) 12,24,12

(b) 24,12,12

(c) 12, 24, 24

(d) 24, 12, 24.

Answer: (c) 12, 24, 24.

Question 15. Appearance of vegetative propagules from the nodes of plants such as sugarcane and ginger is mainly because

(a) nodes are shorter than intemodes

(b) nodes have meristematic cells

(b) nodes have meristematic cells

(c) nodes are located near the soil

(d) nodes have non-photosynthetic cells.

Answer: (b) nodes have meristematic cells.

Question 16. There is no natural death in single celled organisms like Amoeba and bacteria because

(a) they cannot reproduce sexually

(b) they reproduce be binary fission

(c) parental body is distributed among the offspring

(d) they are microscopic.

Answer: (c) parental body is distributed among the offspring.

Question 17. There are various types of reproduction. The type of reproduction adopted by an organism depends on

(a) the habitat and morphology of the organism

(b) morphology of the organism

(c) morphology and physiology of the organism

(d) the organism's habitat, physiology and genetic makeup.

Answer: (d) the organism's habitat, physiology and genetic makeup.

Question 18. Which of the following is a post-fertilization event in flowering plants?

(a) Transfer of pollen grains

(b) Embryo development

(c) Formation of flower

(d) Formation of pollen grains

Answer: (b) Embryo development

Question 19. The number of chromosomes in the shoot tip cells of a maize plant is 20. The number of chromosomes in the microspore mother cells of the same plant shall be

(a) 20

(b) 10

(c) 40

(d) 15

Answer: (a) 20

Question 20. The growth phase of an organism before attaining sexual maturity is referred to as

(a) juvenile phase

(b) vegetative phase

(c) both (a) and (b)

(d) none of these.

Answer: (c) both (a) and (b)

Question 21. Select the monocarpic plant out of the following.

(a) Bamboo

(b) Lite hi

(c) Mango

(d) All of these

Answer: (a) Bamboo

Question 22. Clear cut vegetative, reproductive, and senescent phases cannot be observed in

(a) annual plants

(b) perennial plants

(c) biennial plants

(d) ephemeral plants.

Answer: (b) perennial plants

Question 23. Strobilanthus kunthiana flowers once in

(a) 5 years

(b) 12 years

(c) 20 years

(d) 50 years.

Answer: (b) 12 years

Question 24. Strobilanthus kunthiana differs from bamboo in

(a) being monocarpic

(b) length of juvenile phase

(c) being polycarpic

(d) none of these.

Answer: (b) length of juvenile phase

Question 25. Oestrous cycle is reported in

(a) cows and sheep

(b) humans and monkeys

(c) chimpanzees and gorillas

(d) none of these.

Answer: (a) cows and sheep

Question 26. Which of the following animals show menstrual cycle?

(a) Gorillas and chimpanzees

(b) Monkeys and humans

(c) Orangutans and monkeys

(d) All of these

Answer: (d) All of these

Question 27. Senescent phase of an organism's life span can be recognized by

(a) slow metabolism

(b) cessation of reproduction

(c) decreased immunity

(d) all of these

Answer: (d) all of these

Question 28. If a fungal thallus has both male and female reproductive structures, it will be called

(a) heterothallic

(b) homothallic

(c) dioecious

(d) monoecious

Answer: (b) homothallic

Question 29. Staminate flowers produce

(a) eggs

(b) antherozoids

(c) fruits

(d) all of these

Answer: (b) antherozoids

Question 30. Single-celled animals are said to be immortal because

(a) they grow indefinitely in size

(b) they can tolerate any degree of change in temperature

(c) they can reproduce throughout their life span

(d) they continue to live as their daughter cells.

Answer: (d) they continue to live as their daughter cells.

Question 31. Which of the following has the longest life span?

(a) Banyan tree

(b) tortoise

(c) parrot

(d) Elephant

Answer: (a) Banyan tree

Question 32. Select the option which arranges the given organisms in ascending order of their life span.

(a) Parrot < Crow < Butterfly < Banyan tree

(b) Butterfly < Crow < Parrot < Crocodile

(c) Fruit fly < Crocodile < Parrot < Banyan tree

(d) Parrot < Tortoise < Dog < Crow

Answer: (c) Fruit fly < Crocodile < Parrot < Banyan tree.

Question 33. is a life process that is not essential for an individual's survival but for survival of the species.

(a) Growth

(b) Reproduction

(c) Respiration

(d) Nutrition

Answer: (b) Reproduction

Question 34. 'Clones' are individuals that have exactly the same

(a) Lifespan

(b) physiology

(c) growth rate

(d) genetic makeup.

Answer: (d) genetic makeup.

Question 35. Which one of the following processes results in the formation of clone of bacteria?

(a) Regeneration

(b) Budding

(c) Binary fission

(d) Fragmentation

Answer: (c) Binary fission

Question 36. Asexual reproduction is seen in members of Kingdom

(a) Monera

(b) Plantae

(c) Animalia

(d) All of these.

Answer: (d) All of these.

Question 37. During binary fission in Amoeba which of the following organelles is duplicated?

(a) Plasma membrane

(b) Nucleus

(c) Contractile

(d) All of these

Answer: (b) Nucleus

Question 38. Vegetative propagation is the term used for

(a) sexual reproduction in animals

(b) sexual reproduction in plants

(c) asexual reproduction in animals

(d) asexual reproduction in plants.

Answer: (d) asexual reproduction in plants.

Question 39. Which of the following is not used for vegetative propagation?

(a) Bud

(b) Bulbil

(c) Turion

(d) Antherozoid

Answer: (d) Antherozoid

Question 40. Identify the given organism and find its maximum life span.

(a) Sparrow, 25 years

(b) Crow, 30 years

(c) Crow, 15 years

(d) Eagle, 40 years

Answer: (c) Crow, 15 years

Question 41. Which of the following options shows two plants in which new plantlets arise from the same organ?

(a) Dahlia and ginger

(b) Potato and sweet potato

(c) Dahlia and rose

(d) Potato and sugarcane

Answer: (d) Potato and sugarcane

Question 42. Refer to the given figure and identify X in it.

(a) Offset

(b) Eyes

(c) Runner

(d) Bulb

Answer: (b) Eyes

Question 43. Fleshy buds produced in the axil of leaves, which grow to form new plants when shed and fall on ground, are called

(a) bulbs

(b) bulbils

(c) tubers

(d) offsets.

Answer: (b) bulbils

Question 44. In which one pair, both the plants can be vegetatively propagated by leaf pieces?

(a) Bryophyllum and Kalanchoe

(b) Chrysanthemum and Agave

(c) Agave and Dioscorea

(d) Bryophyllium and Asparagus

Answer: (a) Bryophyllum and Kalanchoe

Question 45. Identify the given vegetative propagule.

(a) Bulb

(b) Runner

(c) Rhizome

(d) Bulbil

Answer: (d) Bulbil

Question 46. If a leaf cell of Agave has x chromosomes, then what will be the number of chromosomes in a cell of its bulbil?

(a) 2 x

(b) x/2

(c) x/4

(d) x

Answer: (d) x

Question 47. Which of the following cannot serve as a vegetative propagule?

(a) A piece of potato tuber with eyes

(b) A middle piece of sugarcane internode

(c) A piece of ginger rhizome

(d) A marginal piece of Bryophyllum leaf

Answer: (b) A middle piece of sugarcane internode

Question 48. Which of the following options correctly identifies artificial and natural methods of vegetative propagation?
Artificial methods – Natural methods

(a) Grafting – Cutting

(b) Layering – Bulbils

(c) Offset – Tissue culture

(d) Tubers – Rhizomes

Answer: (b) Layering – Bulbils

Question 49. Sexual reproduction is considered more beneficial than asexual reproduction because

(a) it is not affected by adverse environmental conditions

(b) fertilization is a chance factor

(c) it rapidly multiplies the population

(d) it assists in evolution by producing variations.

Answer: (d) it assists in evolution by producing variations.

Question 50. Development of new individual from female gamete without fertilization is termed as

(a) syngamy

(b) embryogenesis

(c) oogamy

(d) parthenogenesis.

Answer: (d) parthenogenesis.

Question 51. Fertilization cannot occur in absence of surface water in

(a) Fucus

(b) Funaria

(c) Marsilea

(d) all of these.

Answer: (d) all of these.

Question 52. Spirogyra is a sexually reproducing alga in which vegetative thallus is haploid. In Spirogyra, meiosis

(a) never occurs

(b) occurs at time of gamete production

(c) occurs after fertilization.

(d) occurs during vegetative growth.

Answer: (c) occurs after fertilization

Question 53. Life begins in all sexually reproducing organisms as a

(a) single-celled zygote

(b) double-celled zygote

(c) haploid zygote

(d) haploid gametes.

Answer: (a) single-celled zygote

Question 54. Which of the following is not correct regarding sexual reproduction?

(a) It is usually biparental.

(b) Gametes are always formed.

(c) It is a slow process

(d) It involves only mitosis.

Answer: (a) It is usually biparental.

Question 55. Offsprings of oviparous animals are at greater risk of survival as compared to those of viviparous animals because

(a) proper embryonic care and protection is absent

(b) embryo does not develop completely

(c) progenies are of smaller size

(d) genetic variations do not occur.

Answer: (a) proper embryonic care and protection is absent

Question 56. Deposition of calcareous shell around zygote occurs in

(a) birds and reptiles

(b) birds and mammals

(c) mammals and reptiles

(d) all of these.

Answer: (a) birds and reptiles

Question 57. Select the option which shows viviparous animals only,

(a) Lizard, Turtile

(b) Platypus, Crocodile

(c) Cow, Crocodile

(d) Whale, Mouse

Answer: (d) Whale, Mouse

Question 58. Which of the following animals give birth to young ones?

(a) Ornithorhynchus and Echidna

(b) Macropus and Pteropus

(c) Balaenoptera and Homo sapiens

(d) Both (a) and (c)

Answer: (d) Both (a) and (c)

Question 59. Viviparity is found in

(a) Sharks

(b) lizards

(c) frogs

(d) birds

Answer: (a) Sharks

9 788197 189852

Printed at Repro India Ltd.